THE LIMITS OF DREAM

THE LIMITS OF DREAM

A Scientific Exploration of the Mind/Brain Interface

J. F. Pagel

AMSTERDAM • BOSTON • HEIDELBERG • LONDON • NEW YORK • OXFORD
• PARIS • SAN DIEGO • SAN FRANCISCO • SINGAPORE • SYDNEY • TOKYO
Academic Press is an imprint of Elsevier

Academic Press is an imprint of Elsevier
Linacre House, Jordan Hill, Oxford OX2 8DP, UK
84 Theobald's Road, London WC1X 8RR, UK
30 Corporate Drive, Suite 400, Burlington, MA 01803, USA

First edition 2008

Library of Congress Cataloging in Publication Data
A catalog record for this book is available from the Library of Congress

British Library Cataloguing in Publication Data
A catalogue record for this book is available from the British Library

ISBN: 978-0-12-374215-5

For information on all Academic Press publications
visit our website at http://books.elsevier.com

Typeset by Charon Tec Ltd (A Macmillan Company), Chennai, India
www.charontec.com
Printed and bound in Great Britain by
CPI Antony Rowe, Chippenham and Eastbourne

Transferred to Digital Printing, 2010

Working together to grow
libraries in developing countries

www.elsevier.com | www.bookaid.org | www.sabre.org

ELSEVIER   BOOK AID
              International   Sabre Foundation

For Kathleen

CONTENTS

ACKNOWLEDGMENTS

The author's intellectual journey into sleep began with an award from the charismatic Nazi and architect of our species journey into space, Werner Von Braun, was extended though mentorship with the sleep pioneer Vernon Pegram at UAB, and diverted into dream by the support of the nightmare himself – Ernest Hartmann. The expansion of this work into population and culture could not have been accomplished without the social statistical expertise of Barbara H. Vann and Carol Kwiatkowski. Kathleen Broyles offered expertise and access into the Sundance film labs and the film set. Both John Sayles and Haskel Wexler facilitated the film exploration, allowing me on set, at times even allowing me to hold the camera. Fred Ebert, the biking philosopher, read and critiqued my ruminations into Descartes. Harry Hunt introduced Heidegger. Ross Levin allowed me to peruse his latest work on dream and emotions, critiquing that chapter.

Pastoral English professor Richard Moore introduced me to Eileen Scarry's work and was kind enough to correct both the grammar and wandering tense of the entire text. Margaret Ezekiel acknowledges the Kauai roots of this project with her powerful cover pastel and Kathleen Broyles its paradoxical patterns with her delightful illustrations.

PREFACE

SEARCHING THE BRAIN FOR EVIDENCE OF DREAMING

Since the time of Freud, the body has been ascendant in cognitive science, the mind hidden and apparently described by neuron-to-neuron chemical connections in the central nervous system. The last half of the 20th century presented challenges to scientific understanding and logic that would force a profound change in the older sciences. Mathematics, astronomy, physics, chemistry, genetics, and cellular biology, scientific fields once characterized by simple formulas reflecting concrete fact, transformed into complex interrelated spheres of activity now defined by algorithms of possibility. In contrast, cognitive neuroscience developed in this era, a distinctly new and apparently unique area of science, exemplifying simplicity. Brain activities became amenable to easy understanding when modeled on sequential firing of neurons as modulated by neurochemical transmitters at the synaptic junctions between cells. This simple model has been extraordinarily productive, serving as the basis for medical approaches as diverse as pharmacologically based psychiatry and neurosurgery.

Unfortunately, like most myths that simplicity disappears with closer examination. Simple "truths" of cognitive science flounder. For instance, everyone knows (check the textbooks) that the rapid eye movement (REM) sleep we see occurring every 90 minutes each night is dreaming – right?

Except that almost all research indicates the contrary. REM sleep does not equate with dreaming. REM sleep occurs without dreaming and dreaming occurs without REM sleep. The two cognitive states are doubly dissociable. There is little evidence that an actual physiologic correlate for dreaming exists. Although the REM sleep we see on our electrical recordings of the brain during sleep is a state in which dreaming may occur, dreaming also occurs throughout the other stages of sleep.

Many neuroscientists have built their careers on simple neurotransmitter constructs of CNS functioning. There is good evidence that the neurochemical acetylcholine is an on–off switch for REM sleep, modulated by other neurochemicals including norepinephrine and serotonin. This view has proved to be much too simple even when limited to the electrophysiological process we call REMS rather than it's loosely defined correlate of "dreaming." Each CNS neuron, all hundred billion plus, are likely to utilize more than one neurotransmitter and secondary signaling systems. The neurotransmitters of the CNS have a multiplicity of highly site-specific roles that may result in net changes on behavioral states and cortical arousal that directly contrasts with effects at the microscopic level. Multiple factors and systems are involved, with no single chemical neurotransmitter identified as necessary or sufficient for modulating sleep and wakefulness. The drugs that induce disturbed dreaming and nightmares are not those that affect REM sleep. Drugs that alter REM sleep often have no effects on dreaming. Despite claims to the contrary, we have only a limited, incomplete, and often incorrect knowledge of the neurochemistry of conscious states.

Much of our knowledge of brain functioning comes from pathologic studies and CNS scanning. Positron emission tomography (PET) and other scanning technologies have found areas of the brain that are consistently activated during REM sleep. If REMS is dreaming, these are the areas of brain that result in dream. However, REMS may or may not be dreaming, and there is no way for the investigator to determine whether dreaming in actually occurring when these areas are activated. Pathologic brain damage can affect dream recall. Extensive basal-frontal and parietal damage to the brain can produce a lack of dreaming. There are individuals, however, without a history of brain damage that function normally without ever recalling dreams. Do these individuals actually not dream? Such findings have called into question whether neuropathology and CNS scanning provide any clear evidence to support a brain-based site of origin for dreams.

There are, of course, clear brain based correlates for many areas of cognition. Neuroscientists and clinicians understand the cognitive processes involved in perception and motor activity quite well. We can create artificial constructs of these systems that have more capabilities than biological systems. We are slowly discovering the brain-based sites involved in other areas of cognitive processing. The experience of dreaming includes visual imagery, emotions, and memories. Since dreaming likely utilizes the same neuro-processing systems used for these processes in waking, these components of dream are likely to have clear brain-based correlates.

There are other types of cognitive processing for which there is no clear associated or specific brain activity. Most of these processes are cognitively described as "thinking." When awake, we think. We even think when we are asleep. Dream thought describes the personal presence of the dreamer in the dream. The particular type of associative thought that frames the imagery, emotion, and memory of dreaming is also characteristic of the process of creativity – a process that often incorporates dreams. Associative thought is the framework for imaginative construction, the process through which the perceptually limited and thin images of dream and waking imagery can produce the powerful and sometimes significant experiences of dream that are capable of parodying or surpassing waking perceptually based experience.

The neurobiology of the human CNS is one of the most complex systems ever studied making it tempting to reduce this system into simple constructs. Yet, theories of cognitive function where simple physiological events represent complex cognitive states such as dreaming have outlived their usefulness. Theories are useful if they can be adapted or altered to fit new data. Theories can be harmful if used to misrepresent the extent of knowledge or to limit the breath of study. We find ourselves in an interesting position. Evidence is lacking to support a simple brain equals mind paradigm. Yet, this Monist requirement lies at the basis of current neuroscientific and psychoanalytic theories of mind.

Whether you look from outside/in as a cognitive scientist or from inside/out as a human trying to understand, you are faced with the same unexplainable conundrum. Just where is your mind?

CHAPTER 1

Introduction

The indescribable panic of suffocation, the exhilaration of flight high above the clouds, or the crashing tidal wave pushing houses and trees before it creating the total destruction that marks the end of the world. A dream can be an overwhelming experience containing images such as these, what dream scientists call "contextual" images. But the dream experience is more than a mere image. A dream can include intuitive, creative personal insights that, although they are our own, we have difficulty understanding, even explaining when we wake. Such dreams can be complex beyond descriptive capacity, leaving us confused and disoriented on awakening, questioning both their meaning and import.

There are waking equivalents to the discombobulating sense of reality that one experiences in the dream world, but they are few and far between. Once, I watched a total solar eclipse on the big island of Hawaii. Perceived visually, a solar eclipse is nothing more than the circumference the moon perfectly occluding the sun. Each morning and evening, we experience a similar cosmological experience of the setting of the sun and the rising of the moon. The experience of a solar eclipse differs exponentially. The black line of totality appears in distance, a black line moving across the earth with the speed of a jet plane. It arrives and hits you with the force of a locomotive. You stagger. Strange winds blow. Solar flares – worms of red light – edge past the black disk of the sun. Night birds sing. The earth grows quiet, confused. Time moves slowly. Minutes later, the light of the sun returns, arriving quietly, marking the return of normality. Components of this experience can be explained scientifically. The darkness confuses birds, inducing them to act out twilight behaviors. Abrupt temperature changes induce winds. Internal circadian sensors sensitive to sunlight can be affected by a sudden loss of light. Yet still, the experience of a total eclipse is intensely visceral. One comes away exhausted and disoriented.

Such images in metaphor are the stuff of dreams.

Dreaming is one of the most intensely studied forms of human behavior. Mankind has spent more than 6000 years applying magic, religion, and logic to the study of dreams, and scientists have found reasonable scientific explanations for some components of the dream experience. Techniques have been developed to access and study the process of memory for the dream experience, as well as the visual imagery and emotional context of the dream. Well-described factors affect dream recall frequency. Yet research and writing about dreaming do not necessarily reflect the results of scientific experimentation. Philosophically, dream study is a young science. The literature is replete with grand theories purporting to explain the dream state fully. The role of scientific research has been to support such grand theories: the dream as rapid eye movement (REM) sleep; the dream as a neurochemical construct; the dream as hallucination; the dream as screensaver for the neural computer that is the brain.

The simple constructs of such grand theories often ignore the actual profound experience of dreaming. Many questions remain unanswered. What is dream's relationship to sleep? What is the neural basis of cognitive states? What are the origins of these strange thought patterns, these intense personal stories? What is their purpose? Is there meaning in this complex interplay of thought and image that we call a dream? Could dreams have a function? Is it sensible to use dreams to inform our waking life? Are dreams a part of the creative process?

All of these questions address elusive brain processes that are rarely rationally described, including thought, creativity, and meaning, cognitive functions that remain unexplained by scientific technology and assessment. These cognitive processes are currently beyond the limits of the structured, conceptual logic that we call science. These processes, each a component cognitive process of dream, are part of what some call "mind."

BODY/MIND

In the last 500 years, philosophers and scientists have been trying to solve the problem of the relationship of mind to brain. It has stubbornly resisted their best efforts. The dichotomy of mind and brain forms the basis of our species' attempts to understand ourselves and the world around us. This dichotomy is codified and built into the structures of our fields of knowledge. This mind/brain interface divides the subjective from the objective, conscious thought from non-conscious brain activity, psychiatry from medicine, and art from science. Many of our current sciences began as studies in philosophy, as investigations into the truths and principles of being. Educational accomplishments in the scientific fields were denoted as degrees in philosophy. At one time, the scientific fields of chemistry, physics, mathematics, and medicine were aspects of the study of mind, part of an overall study of philosophy, of what we could not yet understand. Chemistry was alchemy. Math was numerology. Astronomy was astrology. Only later was evidence found to experimentally support their basic laws and tenets. When reproducible tests could confirm speculation, that field of study became what we call science. Once these fields became sciences, their techniques and experiments could be utilized in the study of body and brain.

DREAMS AS WINDOWS

Scientists and philosophers often use dreams in their attempts to understand the basic dichotomy of brain and mind. This approach dates back to Rene Descartes and his insight into the process of logic that we now call the scientific method. On the night of November 10, 1619, Descartes was a soldier fighting for hire in what would come to be known as the One Hundred Year War. Descartes notes in his journal ideas that came to him in a series of dreams:

> And finally, taking into account the fact that the same thoughts we have when we are awake can also come to us when asleep, without any of the later thoughts being true, I resolved to pretend that everything that had ever entered my mind was no more true than the illusions of my dream. For how does one know that the thoughts that come to us in our dreams are more false than the others, given that they are no less vivid or expressed? (Descartes, 1641, p. 17) ... Our dreams ... represent to us various objects in the same way as our exterior senses do ... What truth there is in them (our thoughts) ought infallibly to be found in those we have when awake rather than those we have in our dreams (p. 21).

This is, in part, Descartes' argument for the principle of what we now call Cartesian doubt, the willful suspension of all interpretations of experience that cannot be proved. Descartes asserts that scientific method can be applied to reality but not to dream. Descartes chooses philosophically to assume a perspective that dream is all that we cannot scientifically understand (1637).

For those of us who have come from a cultural standard of belief that brain equals mind, it is difficult to comprehend the possibility that brain may not be mind. It is even harder to realize the compelling power of the Cartesian argument. Yet, the partition of body from mind in philosophy and science prevailed for the next 300 years. As recently as 1958, the German nuclear physicist Werner Heisenberg complained of "the Cartesian partition, which has penetrated so deeply into the human mind during the last century" (1952). In our recent history, the field of cognitive science has been based on the basic general premise that what we call mind is the function of underlying brain activity. It is not uncommon for cognitive scientists to believe that there must be at least a slice of the brain responsible for thought, feeling, spirituality, and dream. Some go so far as to suggest that Descartes' belief that body and mind are different kinds of stuff that interact in the brain should be considered ridiculous by any responsible modern scientist (Pinker, 1997).

Actual evidence for the existence of mind-based brain matter has been difficult to find. The mind = brain correlate has found its strongest support from studies of the association of sleep states with dreaming. A theory called activation–synthesis proposes that all cognitive behaviors, both conscious and non-conscious, reflect the biological and physiological activity occurring in the central nervous system (CNS). This theory, developed at Harvard by Alan Hobson and Robert McCarley, served to dispel the age-old Cartesian dilemma for many cognitive scientists. Brain activity, it asserts, must be the basis for cognitive processes, even those that we call mind. This grand theory needed only experimentally testable scientific evidence to prove that brain activity results in mind processes (McCarley & Hobson, 1975).

Hobson and McCarley looked to the developing science of sleep for evidence to support their hypothesis. The mid-1960s were a romantic time for science. Humans were exploring the near frontiers of space. Electronic monitoring telemetry used to monitor both primates and humans involved in the space program was producing revolutionary data as to norms of physiological functioning over the 24-hour cycles of day and night. Such telemetry forms the basis for the study of sleep utilizing the technique of polysomnography (PSG). A PSG records multiple channels of physiologic telemetry recorded throughout a night of sleep, including brain electrical activity (electroencephalogram, EEG), eye movement (electrooculogram, EOG), voluntary muscle discharge from the legs and face (electromyogram, EMG), heart electrical rhythms (electrocardiogram, EKG), and a variety of sensors for recording breathing during sleep. Some of the first PSGs were recorded at the University of Chicago by a young technician named William Dement. He noted a strange phenomena occurring episodically during the night. Each individual he recorded had periods in which the eyes moved rapidly back and forth. He and his co-workers, Eugene Aserinsky and Nathaniel Kleitman, called the sleep associated with these episodes "rapid eye movement sleep" (REMS). When they awakened their test subjects during this REMS state, most reported dreaming. One dreamer, describing a dream of watching a ping-pong match, led Dement to theorize that these eye movements were actually scanning the visual field of a dream. Dement video recorded an individual who had reportedly never dreamed. Voila! Just as REMS occurred, he was awakened and reported a dream.

Hobson and McCarley used these findings to support their theory of activation–synthesis. If REMS is equivalent to dream sleep, then all we need to do is to look at the monitor in a sleep laboratory. The cognitive activity of dreaming is the electrical EEG activation occurring during REM sleep. Instead of waking an individual to ask whether he or she is dreaming, we can display the EEG waves of the dreaming brain. If REM sleep is occurring, then the person is dreaming. Simple!

BUT, IS REM SLEEP DREAMING?

The belief that REMS is dreaming persists. This is the proof that body equals mind. "REMS is dream" has assumed an irrational mystique. Despite a spate of research books and peer-reviewed papers based on REMS = dream, evidence that the correlate actually exists is strikingly absent. The association between the cognitive experience of dreaming and REM sleep is neither specific nor close. REM sleep is but one of the stages of sleep associated with dreaming. REMS can be understood within the context of sleep without invoking mental phenomena or quasi-conscious processes such as dreaming (Pagel, 2003).

Dreaming is a cognitive state. If we did not experience the recall of sleep-associated images, experience thoughts and emotions upon awakening, dreaming would not exist as a topic for discussion. REM sleep is one of the non-conscious states of sleep that we can electrophysiologically monitor. Dreams are mental activity of sleep that we remember on awakening. Review of current research in the field shows that it is only the rare apologist for activation–synthesis theory that persists in the belief that REMS = dream.

Characteristic of the study of science, theorists, academics, and scientists find at some point during their career that they have built their work on a failed belief, and faced with negative evidence, they strive against odds to maintain the model. It is often the case as well in science that the greatest support for a paradigm shift comes from its defenders. To maintain the REMS = dreaming model, theorists claim that much of the reported cognitive activity occurring during sleep – any thought, feeling, or emotion that is not bizarre and hallucinogenic – is not dreaming and should be excluded from what is defined as dreaming. This other mental activity has no clear name, though authors have called this non-dreaming "mental activity of sleep" (Neilsen, 2003). Hoping that the REMS = dreaming correlate can be saved, some propose that we should radically redefine REM sleep as occurring not only in one stage of sleep (REMS) but also in other stages of sleep as well as throughout waking. They postulate that any bizarre, hallucinatory thought should be considered a dream, whether it occurs in sleep, wakefulness, or in some intermediary state. These apologists for REMS = dreaming argue that such mentation proves that REM sleep must have occurred. In order to preserve the theory that REMS = dreaming, we must redefine both wake and sleep.

There is a major shift occurring in our understanding of the association between sleep and dreaming. REM sleep and dreaming are doubly dissociable states with different physiological mechanisms. Dreaming clearly occurs outside of REM sleep. In all likelihood dreams and REM sleep serve different functional purposes. Scientific methodology can be used to describe the electrophysiological and neurochemical correlates of dreaming (such as REM sleep). But the presence or absence of dreaming cannot be described by our current technology. Dreaming and REM sleep are complex states for which the dreaming = REMS model has proved too simple, and indeed misleading. The paradigm is shifting.

The persistent power of the belief that REMS = dreaming despite the abundance of evidence to the contrary suggests that there is more at play here than a simple change in scientific paradigm. Our modern cultural belief that mind = brain requires scientific evidence. If REMS is not dreaming, it can once again be argued that brain is not mind. A border would have to exist between the two, a border dividing mind from brain – setting limits to the ability of either to describe the other.

THE EVENT HORIZON

Although REMS = dreaming has been generally accepted by the public, the scientific community has been unable to confirm the identification. We are once again confronted by philosophers who have expounded upon their increasing frustration the unproven assumption that mind = brain. Colin McGinn, author and professor at Rutgers University, went so far as to describe the mind–body problem aching in his bones "like an oppressive mist at dawn" (1991):

> "The mystery persists. I think the time has come to admit candidly that we cannot resolve the mystery ... we know that brains are the de facto basis of consciousness," but "we are cut off by our very cognitive constitution from achieving a conception of that natural property of the brain (or of consciousness) that accounts for the psychophysical link" (1989, pp. 349–350).

McGinn goes further to suggest that:

> Everything physical has a purely physical explanation. So the property of consciousness is cognitively closed with respect to the introduction of concepts by means of inference to the best explanation of perceptual data about the brain (1989, p. 358).

Philosophers have described this limiting border existing between body and mind as the "homogeneity constraint." We are locked into whichever side of the subjective–objective divide from which we start. We are never allowed to move from the subjective to the objective or from the objective back to the subjective. This is both a real and a methodological obstacle that makes it impossible to cram the subjective and the objective into the same inferential space (Flanagan, 1992). Nagel says it best:

> We have at present no conception of how a single event or thing could have both physical and phenomenological aspects, or how if it did they might be related (1974, p. 47).

These philosophers suggest that if we are to understand the association between the two states, we must first understand that a closed Cartesian border exists between mind and brain. McGinn goes as far as to insist that the mind/brain border may be absolutely closed to our conscious understanding.

Scientific understanding often requires adherents to move beyond the early simple concepts on which a particular science was based. The older sciences have reached levels of complexity where such limits have had to be faced. For physicists, the simple formulas of classical Newtonian mechanics described both the position and velocity of a particle and set the basis for much of modern technological progress. Yet the newer physics of quantum mechanics require that we now accept that on sub-microscopic levels we cannot know both the position and velocity of a particle at the same time. This is the in-principle impossibility called Heisenberg's Uncertainty Principle that applies as well to the cosmology of maxi-physics. For the mathematician, there is Godel's Incompleteness Theorem that asserts a certain sentence in arithmetic is true though it is impossible to prove it with arithmetic. For cosmologists such confounding limits have become the norm. Time and space may be finite but without any boundary or edge (Hawking, 1993).

Cosmologically, an event horizon describes the boundary of a black hole. At its center is a singularity of infinite density – the end of time for a collapsing star and anything caught within its gravitational pull. At the boundary of the event horizon is a wave front of light just failing to escape the gravitational pull of the collapsed star. Such an event horizon is the perfect border. Everything unfortunate enough to have fallen into this hole in space–time disappears from any reality outside that place. But what if our universe has not really lost the information that has fallen into the black hole; that it is simply lost in a region of space cut off from the rest of the universe, where it will be compressed and reduced to the essential particles of cosmological creation? What if information can somehow sneak back across that border?

Stephen Hawking describes how black holes radiate and shrink, how the thickness of the border around a black hole is proportional to the size of the black hole itself; how sub-atomic particles are rapidly leaking out of smaller holes, permitting data once lost to re-enter our universe (1992). Physicists inspired by Hawking's findings and armed with Einstein's equations have begun further exploration into the properties of a black hole's core. Their findings have the feel of science fiction – that the singularity at a black hole's core could be the gateway to another universe or back into this universe through other holes in space–time (Green, 2003; Hawking, 1993). Some scientists theorize that we would experience nothing at all if we were to cross the event horizon of a black hole (Luminet, 1992). Others view the exploration in a different light. There is a reason Hawking was too excited to sleep after his realization of the particles and therefore information could leak across the event horizon of a black hole. No other border has ever been perfect. No one area of a nation-state, or of conceptual mathematics, or physics, or cognition and even space–time can be completely separated from another. Someone or something is always sneaking across the border.

Exploration of the border between mind and brain could be a dry philosophic or intellectual exercise. Yet borders tend to be strange and wondrous places. For this mind/brain border, these are exciting times. In scientific thought as in collapsed stars, old paradigms are fading, even those that had been generally accepted as true. No border is perfect, not even the event horizon's region of space–time from which in theory, escape is not possible. We can, in fact, detect radiation produced by elementary particles that are quantum mechanically tunneling past the event horizon of the black hole. Ideas, like newly detected wave fronts, are crossing borders that were presumed uncrossable, once again being considered and coming into the light.

Dreams can be considered leaks in the border between body and mind. The dream leaking from sleep, a state cutoff from waking experience, has the astonishing property of appearing as if it has originated from some external reality. Descartes pointed out that dreams are a window into the understanding of mind. Freud pointed out that from the perspective of mind, dreams can be a window into the functioning of the brain. Attempting to assume the perspective of mind, dreams describe what can be known of the brain. From the perspective of brain science, dreams describe what is known of mind. In our attempts to understand the CNS, the most complex system that our science has ever tried to understand, dreams can serve as a window though the event horizon existing between the mind and the brain. Think back to contextual images from dreaming: crashing buildings, the unbased and totally free sense of falling, the unfathomable darkness, or the depth and weight of immeasurable water. Perhaps these are nothing more than tiny leaks through the event horizon between mind and brain. Flickering images on the walls of Plato's cave made by a singularity that we cannot see (Fig. 1.1).

FIGURE 1.1 Plato's cave.

THE LIMITS OF DREAM

This book is a scientific exploration of the limits of our understanding of dream. The approach that we will take is not nearly as simple as REMS = dream, rather this book will focus on what we currently know of the human CNS, examining the basic sciences of neurochemistry, neuroanatomy, and electrophysiology as these sciences apply to dream. We will reach beyond basic science to examine the approaches used by the cognitive sciences study dreaming, including research into memory, emotion, and perception. We will look at how these cognitive operations are integrated into the related processes of visual imagery and dream. Building on what is known of intrapersonal CNS processing, we will step outside the physical body to explore the status of cognitive processing in artificially intelligent systems as well as the creation of artificial dreams and the use of dreams in filmmaking, art, and story. We will look at the role of dreaming in creative process as well as in creative "madness," examining the limits of knowledge in each of these areas of scientific exploration using the study of dream as a window to explore the border between body and mind.

It is possible that despite initial miscues such a REMS = dream, scientific technology is at the brink of describing a unity between body and mind. However, the contention of this book is that we know far less about both the brain and the mind than pretty PET scan pictures and popular press pronouncements of breakthroughs would have us believe. This book will assume a Cartesian perspective examining the mind/brain interface from each side of the mind/brain border rather than viewing mind and brain as a unitary whole. We will use neural and cognitive science, complexity theory, and the sciences of artificial intelligence to explore the limits of the capacity of neural network systems to produce aspects of mind. We will utilize psychoanalytic approaches as a tool to allow us to attempt to understand the brain/mind interface from the perspective of mind. From each perspective we will push to the limits of what we understand about dream to explore the border that exists between body and mind. We will find that using such a Cartesian perspective, based more on evidence than belief, produces insights and understanding of both body and mind that remain unavailable to the Monist attempting to dig in and protect the more simple unitary theories of mind = brain.

The assumption of a Cartesian perspective of mind/brain might initially appear to be a more complicated and difficult perspective. Yet in assuming this perspective we are freed to explore the science rather than the theory of this border. Both waking and sleeping consciousness are global states developing out of the complex system of the CNS of the human brain. You can look for evidence of the mind in this system using the technological approaches that we have developed to study neural and cognitive science. In this book, we will follow each of these neuroscientific approaches to the limits of our current knowledge. In some of these areas, we will find that brain functioning can explain manifestations of what is called mind. In other areas, we will come up against a series of apparently uncrossable limits between the brain and the mind that are more difficult to rationally describe using a logical cognitive approach. The border between mind and brain can appear like an eclipse horizon, as an apparently uncrossable barrier. Describing this interface is something like trying to explain my visceral experience of a solar eclipse. In order to approach this border, we are going to explore the scientific limits of dream. We will explore the limits of this mental state, the dreams that arrive as leaks tunneling through the border between mind and brain. We will be traveling to a place that can be both beautiful and frightening, exploring the limits of what we as human beings are able to understand.

Definitions and the Search for Truth

Dreaming, mind, and creativity are the focal concepts of this book. But these concepts have alternative, sometimes contradictory meanings for different people based on their background, prejudice, and training. There has been a general tendency for scientists to avoid defining shared cognitive states. Freud's inexact prose and mistrust of exact definitions permeates the field of cognitive science (Mahoney, 1987). Dreaming, mind, and creativity are just some of the poorly defined cognitive states that include consciousness, sleepiness, happiness, fatigue, ecstasy, hope, depression, madness, and despair among others. Over 400 years ago Rene Descartes pointed out, "Now the principal and most frequent error that can be found in judgments consists in the fact that I judge that the ideas, which are in me, are similar to, or in conformity with certain things outside me" (1641). In order to interactively discuss or study a topic, we must first define that which we have decided to talk about.

A definition is a formal statement of meaning. Conceptually, for the purpose of communication, it seems self-evident that it would be best if each term had a definition of essence on which we could all agree. However, in the case of a shared cognitive state such as dream, many definitions are available. Each of us is likely to have our own concept of meaning. We find ourselves unable to agree on an essence of the truth that is the definition of such a concept. Instead we peel away our preconceptions looking rather for the ancient Greek conception of truth in which something that is true is that which is "unhidden." We can create a definition that sets up or acknowledges the limits of accepted usage. In the journey required to approach such a definition, we will find ourselves exploring the basis of our cognitive concept of mind, dream, and creativity. We are not searching just for one truth, but for a series of interconnected and unhidden truths that leave each of us understanding better our own meanings.

CHAPTER 2

Mind/Plato's Cave

Picture people dwelling in an underground chamber like a cave, with a long entrance open to the light on its entire width. In this chamber people are shackled at their legs and necks from childhood, so that they remain in the same spot, and look only at what is in front of them, at what is present before them. Because of their shackles they are unable to turn their heads. However, light reaches them from behind, from a fire burning up higher and at a distance. Between the fire and the prisoners, behind their backs, runs a path along which a small wall has been built, like the screen at a puppet shows between the exhibitors and their audience, and above which they, the puppeteers, show their artistry.

I see, he says.

Imagine further that there are people carrying all sorts of things along behind the screen, projecting above it, including figures of men and animals made of stone and wood, and all sorts of other man-made artifacts. Naturally, some of these people would be talking among themselves, and others would be silent.

A peculiar picture you have drawn, and peculiar prisoners!
They are very much like us! Now tell me, do you think such people could see anything, whether on their own account or with the help of their fellows, except the shadows thrown by the fire on the wall of the cave opposite them?

(Plato's Cave – The First Stage, from
The Republic Book VII, 514a–517a)

It can be sobering to confront the complexity and modern resonance of the ancient Greek philosopher. These days we are often asking for simple answers to complicated questions. The modern philosopher/scientist is more likely to quote an 8th century monk than Plato. The monk Occam proposed that when assessing a series of potential solutions to a religious question, the simplest solution is most likely the correct solution. This truism that we call *Occam's Razor* was first adopted by theologians before being appropriated by mathematicians and scientists. In our complicated modern world, we have come to demand simple answers, applying *Occam's Razor* to the seemingly insoluble social, political, and financial problems of our daily lives.

What is the mind? This is a question more difficult than nearly all others. Mind as a philosophical concept has had many meanings, with those meanings changing as each generation applies its unique cultural, social, religious, and logical mores into defining the concept. Studies into the relationship of body and mind have developed into complete philosophical traditions. There are periods of pervasive cultural and religious belief in which the definition of mind has seemed so self-evident as to not require definition even for philosophers, but within a generation, the argument of body versus mind would resurface as definitions and perspectives of both body and mind changed based on changing cultural mores or technologies. In each generation, there are also those who maintain the perspectives on mind coming from older traditions. Our era maintains the tradition of avoiding difficult definitions, presuming that what one individual is referring to when mentioning the concept of "mind" is the same as for all others. Philosophers faced with conceptual confusion may resort to a definition of mind based on customary usage (Churchland, 1986). Today it is not unusual for book authors to include "mind" in their title and then never define or specifically address their sense of the concept.

Studies of mind are often philosophic treatises. The philosophic overview included in this book is limited, and readers with a primary interest in philosophic tradition will likely not find their views challenged by its arguments. This book approaches the borders of the mind from the scientific perspective of a modern cognitive neuroscientist reflecting his approaches to this issue from within this most recent era's conception of mind.

Conceptually, discussions of differences between mind and body date back to early civilization and are to be found in our first religious texts:

> Jesus said, "If the flesh came into being because of spirit, it is a marvel, but if spirit came into being because of the body, it is a marvel of marvels. Yet I marvel how great wealth has come to dwell in this poverty (The Gospel of Thomas: The Nag Hammadi Text [29] approximately 100 AD).

In this quote attributed to Jesus the ancient skeptic, there is the suggestion that mind – here called spirit – is all that is not flesh. From some religious points of view, the concept of the separation of body and mind has been interpreted to mean that all that is not flesh can be attributed to coming from outside the individual. In other words, mind is god or god-like, and flesh is human. There are still many who accept this religious Cartesian conception of the separation of body from a greater mind. Other spiritually based concepts of mind include the interior concept of the spiritual mind as "soul" with both body and mind existing within the individual independent of any outside spiritual source. Both the interior concept of mind as soul, and the philosophic perspective of the external greater mind or spirit are religious perspectives of mind. Both concepts are based on belief rather than testable proof.

DESCARTES

The scientifically based discussion of mind began in the 1600s with Rene Descartes. He stated, "Since I do not believe that what I seem to sense in my dreams comes from things external to me, I did not see any reason why I should have these beliefs about things I seem to sense while I am awake (Descartes, 1641)."

Descartes concluded that in perceptual isolation, man was at the center of his own universe. "But I know now with certainty that I am, and at the same time it could happen that all these images – and generally, everything that pertains to the nature of the body are nothing but dreams … But what then am I? A thing that thinks. What is that? A thing that doubts, understands, affirms, denies, wills, refuses, and which also imagines and senses" (ibid., p. 68). Working from that original description of "self," Descartes went on to differentiate his dreaming world from his reality without having to rely on the sometimes undependable evidence of his senses. "I should no longer fear lest those things that are daily shown me by my senses are false; rather the hyperbolic doubts of the last few days ought to be rejected as worthy of derision – especially the principal doubt regarding sleep, which I did not distinguish from being awake. For I now notice that a very great difference exists between these two; dreams are never joined with all the other actions of life by the memory, as is the case with those actions that occur when one is awake" (ibid., p. 100).

In his "Discourse on Method" and later his "First Meditation," Descartes left a description of his attempts to distinguish the attributes of dreaming from an outside interactive perceptual world that was not dreaming. Descartes suggested the possibility that an evil demon produced sense data in our minds but made sure that nothing ever corresponded to them, so that all of our sensory experience was a delusion. This was a difficult argument to refute then, and a difficult argument to refute philosophically now. His original skeptical argument still has considerable resonance among modern philosophers. The argument that our "reality could be dreaming" can still not be clearly refuted (Blumenfield & Blumenfield, 1978; Schouls, 2000).

There is little question that our perceptual concept of the external world that seems so real to each of us is but one of many possible representations of external reality. What we perceive is always sense data, not the real thing. We live in a world of naïve realism where our consciousness acquaints us with its own contents, a play of images upon a mental screen (McGinn, 2002).

Descartes developed a technique for studying these differences he perceived between body and mind by cultivating for himself a form of psychological and perceptual isolation (Vrooman, 1970). That technique developed into what we now call the scientific method. "I learned to believe nothing very firmly concerning what I had been persuaded to believe only by example and custom" (Descartes, 1637). This was to be the first step of his scientific method: (1) never accept anything as true that is not known to be so. Descartes' method had a rigorous and formal logic that also included steps: (2) divide the problem into as many component parts as possible; (3) build from the smallest of what is known; and (4) keep complete records for review. Descartes' first philosophic use of this method was in differentiating dreaming from waking reality. In applying scientific method to differentiate waking reality from dreaming Descartes created a basic structural paradox that still affects the scientific study of dreaming. If a state can be scientifically studied, measured, classified, or logically proven to exist, it is not what Descartes defines as a dream.

This concept of the separation between body and mind is called Cartesian dualism. Cartesianism was from the first attacked as subversive to religion. Descartes, the consummate philosopher, confronted the powerful Roman Catholic Church by redirecting and arguing the opposite side of his own argument. He proposed that the rules and logic that could be experimentally tested with the logic of the new sciences were evidence for the existence of God. He convinced his friend the Jesuit Cardinal Berulle that the evidence for the existence of God was to be found in the scientific details of the natural world. From this perspective the mind is what we know. All else is body. While Galileo fretted under house arrest, Descartes obtained though the Jesuits the support of the Roman Catholic Church, and went off to Sweden to join Queen Christina's court. Descartes the philosopher proposed for his own tombstone a very concrete and politically appropriate epitaph: "He who hid well, lived well" (based on Ovids Trista: "Bene qui latiut, bene vixit").

LIMITS OF THE CARTESIAN PERSPECTIVE

In our time, it seems beyond question that mental events occur as a result of brain-based neural processes. This problem of mental causation is not supported by the Cartesian mind/body belief system in which body and mind are considered as separate. Most of us take it for granted that our mental events have physical effects. We see a commercial for pizza on TV, decide to move our ponderous physical bodies down the stairs and across the street to the restaurant, where we initiate and complete the process of ingestion. We consider that this happens because our beliefs and desires have caused the appropriate controlling neurons and neural systems to fire in our brains, initiating and then completing the physical processes that we require. How can this happen unless our beliefs and desires (our mind) are part of the physical happenings of the brain? It seems obvious that mental processes must simply be physical processes occurring in the brain. Perhaps the mind is hidden in the frontal cortex, the hippocampus, the limbic system, or, as Descartes suspected, in the pineal gland.

Philosophically, this anti-Cartesian perspective that mind is consistently based on physical processes in the brain leads us to make several assumptions:

1. *Mind–body supervenience*: The mental supervenes on the physical in that any two things exactly alike in all physical properties must have the same mental properties (i.e., there is no mental difference without a physical difference).
2. *The anti-Cartesian principle*: Nothing can have a mental property without having a physical property (i.e., nothing that is purely mental can exist).
3. *Mind–body dependence*: Mental properties depend on physical properties. The psychological character of a being is wholly dependent on its physical character.

These basic "Monist" principles result from a logical approach to the belief that body and mind are unitary. If you believe these principles, you can consider yourself a Monist. Rejecting them at any level suggests that you have become a Cartesian. Contemporary discussions of body/mind are generally conducted from within this Monist framework with these three principles of the correlation between mind and body assumed or taken for granted (Kim, 1998).

WHAT IS MIND?

Definitions

If your belief is that all cognitive process can be scientifically explained, that mental processes are identical to neural processes, you believe that body = mind. This mind–brain "identity" theory of "Monism" has in the last 50 years become the most influential perspective of mind. Based on the unitary perspectives of Monism, mind has become defined as the set of cognitive processes based on the brain activity (body) (Place, 2004):

> Mind – (in a human or other conscious being) the element, part, substance, or process that reasons, thinks, wills, perceives, judges, etc.: *the processes of the human mind.* Webster's New Universal Unabridged Dictionary (1996)

This is currently the most accepted definition for mind. However, like many definitions, this descriptive definition reflects the systems of belief that are most widely accepted at this point in time. Some philosophers include sensation, perception, memory, learning, and action as parts of the mind. Based on such a perspective, mind includes the portions of these cognitive processes whenever they occur without waking perceptual input (Damasio, 1994). Perception, memory, learning, and motor activity are brain functions that we understand very well. If mind is nothing more than these processes occurring when we are cut off from perceptual input when we are asleep, then it can be said that we understand mind quite well. The definition of mind becomes the cognitive integration of perception. Ergo, whatever occurs in sleeping and dreaming is conceptually mind. Taken to a Monistic extreme: "... 'the mind' includes such basic activity as the routine firing of neurons that keep us breathing" (Hobson, 1994).

Just as with our current approach to rapid eye movement sleep (REMS) = dreaming, this simplified approach has considerable advantage to the researcher. Since "mind" must be present in primates and other animals that sleep, experience emotions, and have REM sleep, mind can be studied in animal models (Kim, 1998). It is more culturally acceptable to experimentally induce lesions and insert electrodes in such animal subjects. In humans, we are limited to studying how central nervous system (CNS) disturbance and trauma induce specific abnormalities in function.

From this reductive viewpoint, the function of the human CNS is to integrate perception with memory in order to produce motor actions. Computers can be utilized to integrate multiple perceptual inputs with memory systems. If CNS function for both body and mind is this simple, the development of artificial intelligence is right around the corner. But what about the process of thinking, what about thought? What about feelings, esthetics, and spirituality? What about creativity? These are messy topics where we know far less about what is occurring in the CNS and where. It would make for easier research if these poorly defined processes, not amenable to research in animal models, would just go away. But these are the processes that are most clearly mind.

Psychoanalytic Approaches

If brain is all plumbing and wiring, mind should be most apparent to the neuroanatomists and electrophysiologist. However, much of our understanding of mind has come from workers in the various schools of psychoanalysis. It was Freud's insight that logical scientific approaches could be used to study mind. If the processes of mind are experienced and reported, these processes could be described by an observer. The Freudian psychoanalyst is such a trained observer. Trained as a neurologist, Freud was searching for associations between types of psychoanalytic process and psychiatric disease, between psychoanalytic constructs of mind process and brain structures. Freud postulated that if brain = mind, we should be able to find correlates between mind-based psychoanalytic reports and brain functioning.

It is an uncomfortable truth in the field of psychoanalytic study that evidence has been difficult to find correlating diagnoses and therapy based on psychoanalysis with biological disease states. Neuropathologic studies and brain scanning techniques have not demonstrated sites for the expected neural correlates of psychological structures in either psychiatric disease states or normal neural processing. Unless, of course, dreaming = REMS. Fifty years ago, Jouvet discovered in his cat studies that REM sleep required output from the brain stem (1959). Here was the "smoking gun," the evidence that Freud's psychoanalytic constructs had been validated. If REMS was dreaming, here were dreams, constructs of the "id" rising out of brain stem portions of that ancient reptilian brain that we share with all animals. This apparent substantiation of psychoanalytic theory is in part the reason that REMS = dreaming was accepted as gospel based on so little scientific evidence. Activation–synthesis (the Hobson–McCarley hypothesis) at its most basic level can be stated as the claim that our behaviors both conscious and nonconscious reflect biological and physiological activity occurring in the CNS. Perhaps the most important and theory of cognitive science since Freud and the discovery of neurotransmission, activation–synthesis has become so influential and pervasive as to almost single handedly dispel the confusion of the age-old Cartesian dilemma. There could be, it appeared, no longer any question as to whether body equals mind, for you needed only to look at the monitor where the electroencephalogram (EEG) activity marking the onset of REM sleep to see the phenomena of dreaming.

Activation–synthesis has become the most elaborately developed Monist model of brain functioning. The classic model of activation–synthesis describes neural activation interacting with synthesis to achieve mind-based consciousness. The scientific evidence that supports the existence of brain-based correlates of psychoanalysis is built primarily on this model. Yet REM sleep and dreaming are, nevertheless, doubly dissociable. Dreaming occurs outside REM sleep and REM sleep without dreaming. If dreaming and REMS are doubly dissociable states, there remains little if any evidence for a biologic basis for the claims of psychoanalysis (Fig. 2.1) (Freud, 1923).

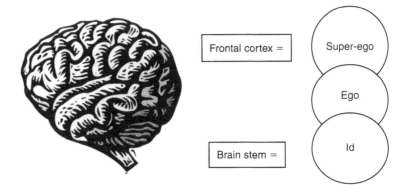

FIGURE 2.1 Super-ego, ego, id, and postulated brain-based correlates.

That is not to say that the years of research and the reams of writing on psychoanalysis are wrong and wasted avenues of endeavor. Psychoanalysis has been one of the few attempts to utilized scientific approaches to study reported phenomena of the mind. If we accept that a dichotomy may exist between body and mind, there is no longer the requirement that the seat of the id is in the brain stem or the super-ego in the forebrain. From the Cartesian perspective, the extensive literature and research into psychoanalysis becomes a treasure trove we can use in our attempts to understand mind.

Psychoanalysis provides approaches that can be used logically and scientifically to study mind. Free association and dream interpretation are techniques that give us information about the mind. The varied psychoanalytic approaches to interpreting that information, Freud's super-ego, ego, and id, Jung's archetypes, Lacan's "Other," are structured approaches that may reflect how the mind is structured, independent of biologic structures. Psychoanalysis has in fact been most useful when applied outside the arena of the diagnosis and treatment of psychiatric illness. Psychoanalytic techniques have proven quite useful in analyzing and studying creativity, particularly in the evaluation of creative process and creative product. While psychotherapy is rarely used today in the treatment of disease, Freudian and post-Freudian approaches have become the "orthodoxy" of the fields of art criticism and film study.

Alternative Perspectives

The consciousness that we experience is quite different from the neural processes that we can observe occurring in the brain. There seems little argument that mind processes include such loosely defined attributes as thought, emotion, reasoning, creativity, belief, and consciousness. It is currently unclear what neural activity is involved in producing these cognitive processes.

The definition of mind has changed with time. But from Descartes, from the beginnings of philosophical logic, the concept of "mind" has included all that we cannot scientifically understand (Descartes, 1637). Since this book addresses both body and mind, we will approach this border from both sides of the divide utilizing a Cartesian definition that mind is not body. Mind includes those cognitive processes that we cannot measure or define with our technology. Mind processes include at least thought, reasoning, creativity, spirituality, belief, and consciousness. These processes are beyond the border of what we can describe with CNS scanning techniques such as positron emission tomography (PET), functional magnetic resonance imaging (f-MRI), and single photon emission computer tomography (SPECT), all powerful tools for the visual analysis of the brain.

We are perceptually bound within Plato's Cave. Mind lies somewhere beyond a flickering reflection on the wall of our cave, a reflection that we believe to be our reality. The mind is hidden somewhere beyond the evidence of our perceptions, somewhere beyond an event horizon that we cannot penetrate with our scientific probes. In this book, we will use our scientific and cognitive understanding of dreaming as that probe, using what we understand of the biology of dreams to explore the border of the mind. Mind, the fire that casts shadows. Mind, the singularity.

CHAPTER 3

A Dream Can Be Gazpacho

Definitions   22

The vision of dreams is this against that, the likeness of a face confronting a face.

(Dead Sea Scrolls, 180 BC)

Over 6000 years ago in Ladak, a city of ancient Samaria, King Gudea had one of his dreams inscribed upon a clay coda. The coda tells the story of how God came to him in a dream and gave him instructions of how to position his temple according to cardinal points. Archeologists have found multiple imprints of that dream inscribed upon fire hardened rolls of soft river clay that had lain covered by the detritus of civilization. One of the codas of this dream is housed in the Istanbul Archaeological Museum at Topiki Palace today, recording one of our species' first documented attempts at written language.

The dreams of kings, prophets, and priests were sometimes considered to be messages from god. These dreams became the stuff of religion, scripted into walls of Egyptian Tombs, the Torah, the Bhagavad-Gita, the Bible, and the Koran. But of course not every dream was a message from god. And it was not just the kings and priests who were dreaming. Children, atheists, criminals, the mentally ill, and even women dream, and these dreams were usually not received as messages from god. For the founders of states and religions that were based at least in part on dreams, the first studies of dreams were designed to determine the true and important dreams so that they could be divined from the false.

For the Egyptians and early Greeks, this attempt to divine "true dreams" from false lies at the basis of their logic, philosophy, and science. Dream analysis became the basis for the science of medicine. In 3rd century BC Greece, physicians and priests required their patients to sleep at the foot of the god in the temple of Aescupales "incubating" a dream that was to be reported the next morning. The physician's task was to interpret that dream and prescribe a cure based on his interpretation of the dream's significance. Today the Aescupales Temples are littered with stone tablets describing these recorded dreams and the subsequent attempts at cure. These are historical examples of some of our first recorded experiments, in which symptoms, dreams, and attempts at treatment, both successes and failures, are recorded. This is the first example of early scientific method as applied to medicine. There are surgical instruments, as well, at the temple, the unusually creative products of what were probably very interesting dreams. The rare successes of the physician–priests that were recorded include the surgical removal of the bladder stone and the trepanome – a hole bored through the skull to relieve cranial pressure – for the treatment of debilitating headaches resulting from intracranial bleeding. Later and even into our days of modern medical technology, these same approaches may be used on patients with similar symptoms.

DEFINITIONS

Dreams are … excursions in my own mind.

(Samuel Coleridge, 1836/1912)

The dictionary definition for "dream" is Aristotelian – mental activity occurring during sleep (Random House, 1987). Other historic definitions have filtered to dictionaries from literature. Baroa, in 1670, recorded, "What is life? A madness. What is life? An illusion, a shadow, a story. And the greatest good is little enough: for all life is a dream, and dreams themselves are only dreams." Psychoanalytic definitions have sometimes assumed such a literary beauty: "The dream is a small hidden door in the deepest and most intimate sanctum of the soul, which opens into that primeval cosmic night that was soul long before there was conscious ego and will be soul for beyond what a conscious ego could ever reach" (Jung, 1934/1965). Biologic definitions may have less literary beauty but can have the profound structure of an architectural creation: "Dreams are the manifestation in consciousness of the operation of the subconscious mind" (Hadfield, 1954). Psychologists construct their definitions with boxes of logic describing the milieu in which dreaming can exist: "Dreaming seems to represent the mind's mode of conscious organization in any circumstance in which external stimulation is reduced or occluded and voluntary control of consciousness is relinquished, while cognitive function is maintained" (Foulks, 1985). More recently there have been attempts at encompassing definitions that would be able to include many of the alternative perspectives of dreaming under a single definition: "By dreaming we refer to any image, thought or feeling, attributed by the dreamer to a pre-awakening state" (Moffit, 1993).

Since almost all of us experience dreaming, we each have our own idea of what a dream is. Yet the awareness, experience, and even the concept of what is and is not a dream are different for different dreamers. For those of us who formally study dream, the meaning, content, allusions, and associations of dream (undefined) can become the focus of intense discussion and publication. These discussions cross broad spectrums of knowledge including topics as diverse as rapid eye movement (REM) sleep, personal motivation, neuroanatomy, film, hallucinations, literature, neuropharmacologic effects of medication, real estate, the unconscious mind, and anthropology. Dreaming (undefined) is often the topic for discussion. Just Google "dream" and see what appears on your computer screen.

It is difficult to know how to approach a process that is so individually defined. Of 100 articles published since 1991 in "*Dreaming*," the journal for the scientific study of dreaming, only 14% include a working definition for dream (Table 3.1) (Pagel et al., 2001). The remaining 86% presume that the reader shares

TABLE 3.1 Working Definitions of Dream and Nightmare from the Journal "*Dreaming*" – March 1991–August 1998

Definition	Field of Study	Topic of Paper	Author	Reference
1. Dream – the expression of fragmentary activity of the brain.	Psychiatry/ Philosophy	Dream Content	G. Globus	1991; 1(1): 27/38
2. Dream – instantiations of a unifying concept.				
3. Dreams are private mental acts which have never been recorded during their actual occurrence.	Anthropology	Anthropology of Dreaming	B. Tedlock	1991; 1(2): 178
4. Dreaming – a product of neurophysiological release during rapid eye movement sleep.	Psychology	Literature/Science	H. Hunt	1991; 1(3): 235
5. A dream is a reflection of our disorganized thoughts … nothing more than disconnected meaningless fragments of what we have been thinking (Gogol).	Literature	Dreams in Russian Literature	L. Visson	1992; 2(3): 181
6. A dream is a poem.	Psychology	Metaphorical Analysis	W. Webb	1992; 2(3): 191
7. Dreaming – "primary" mental functioning characterized by sleeping thought.	Sleep Medicine and Sociology	Dream Effects on Behavior	J. Pagel and B. Vann	1992; 2(4): 229
8. A dream is a dream is a dream.	Literature	Meaning of Dreams	B. States	1992; 2(4): 261
9. Dreams represent the higher brain's efforts to make sense of the haphazard signals generated by the lower brain.	Psychology	Dream Interpretation	C. Hill et al.	1993; 3(4): 269
10. Nightmare – a vivid and terrifying nocturnal episode in which the dreamer is abruptly waken from sleep.	Psychology	Nightmare Subjects	R. Levin	1994; 4(2): 127
11. Dream – a series of images that occur during sleep … often reported in narrative form.	Psychology	Teaching Dreamwork	S. Krippner et al.	1994; 4(4): 215
12. Dreams are human experiences that remain private until they are shared through social, usually discursive, interactions.	Psychology	Dream Sharing	H. Stefanakis	1995; 5(2): 95–104
13. Dreams, which occur while sleeping, may be defined as subjective cognitive experiences with complex mental images that change and progress over time.	Psychology Gerontology	Age Changes in Dream Recall	L. Giambra et al.	1996; 6(1): 17
14. By definition, dreams are experiences arising during sleep.	Sleep Medicine	CNS Hallucinations	Mahowald et al.	1998; 8(2): 89–102

Source: Pagel et al. (2001).

the authors' unstated definition. And that definition is only one among a bewildering diversity of definitions for dream that depend on, at least in part, the area of knowledge and type of training involved in the authors' field of study. What is a dream? What it is to a sleep medicine physician – mental activity occurring during sleep – is very different from what it is for a student of psychoanalysis – bizarre or hallucinatory mutation occurring during sleep and other states of consciousness. As Freud pointed out, "In spite of many thousands of years of effort, the scientific understanding of dreams has made very little advance [defining] little or nothing that touches upon the essential nature of dreams" (Freud, 1907/1953). The scientific study of dreaming has not progressed at the rate of other cognitive sciences. This lack of progress toward an understanding of dreaming is due, at least in part, to the lack of a definition for "dream."

The current standard psychiatric definition of dream is "bizarre or hallucinatory mental activity occurring in the CNS" (DSM-IV, 1994). Such dreaming is often considered to occur during a continuum that extends through stages of sleep and awake (Hartmann, 1998). Based on this definition, some authors have postulated that dreams are hallucinations very similar to the perceptual auditory or visual experiences of individuals in psychotic states (Hobson, 1988). The field of sleep medicine, still in its adolescence, defines dreaming as mental activity occurring during sleep. A sleep physician will ask his patient upon awakening whether images, feelings, or emotions are recalled from sleep. Psychoanalytically defined dreams can occur in sleep or wake and include bizarre hallucinatory content. Sleep medicine defined dreams occur in sleep and can include a wide variety of mental content that may be no more than a color, a feeling, or the awareness that a dream has occurred. Since these two definitions are in many ways mutually exclusive, the psychoanalyst and the sleep physician are often addressing totally different topics when they discuss dreaming.

Other fields such as psychology tend to avoid defining dream, concentrating instead on defining methodology and their study populations, and often not asking their subjects to describe their meanings for dream. Traditional poetic and dramatic writings, as well, tend not to define a dream, choosing rather to evoke its attributes, especially the qualities of illusion and strangeness. In the popular culture, the most generally accepted definition for a dream is loosely Freudian: dream marriages, dream homes, the projected images of conscious wish fulfillment.

So what is a working definition for dream? The experts selected from a cross-section of the fields that study dream have agreed that a single definition for dreaming is impossible (Pagel et al., 2001). These experts met repeatedly before a consensus was reached to develop a classification system that could be used to organize the multiple and sometimes conflicting definitions for dream (Fig. 3.1). The group could agree that the dreaming process has at least three characteristics:

1. *Dreams occur in either awake or sleep:* Dream-like states occur during sleep onset (hypnagogic phenomenon), awakening (hypnopompic phenomenon), meditation, drug induced states (anesthetics, sedative/hypnotics), day dreaming, associated with hallucinations, and during looser, less consciously structured waking thought. Dreaming has been proposed to be a continuum extending through the stages of sleep, to sleep onset, and day dreaming, to focused waking thought (Hartmann, 1998).

(a)	Sleep	Sleep onset	Dream-like states	Routine waking	Alert wake
----------*----------------*----------------------*----------------------------*---------------------------*------------					

(b) No recall	Recall	Content	Associative content	Written report	Behavioral effect
----------*----------------*------------*----------------------*----------------------------*--------------------*------------					

(c) Awareness of dreaming	Day reflective	Imagery	Narrative	Illogical thought	Bizarre hallucinatory
-------------*------------------------------*------------- *-----------*-------------*----------------------*-----------					

FIGURE 3.1 **Definitions for dreaming – a classification system paradigm.** A definition of dream has three characteristic continua (a) wake/sleep, (b) recall, and (c) content. Source: Pagel et al. (2001).

2. *Dreams are generally recalled and brought to awake conscious awareness:* Awake dream reports are generally accepted as the evidence as to whether a dream has occurred. Dreams may occur independent of any subsequent waking recall or reporting. Examples include dreams of infancy, and animal dreams in which recall is limited by reporting capability; and dreams without initial recall that are remembered later. This continuum describing recall extends from no recall, to awareness of dreaming, verbal reports, associative facilitation of reports, written dreams, to the incorporation of the dream into behavior.

3. *Dreams have a describable content:* Dreams have content that consistently varies from that of waking thought, in that outside observers can distinguish between transcribed dreams and awake thoughts. The characteristic content of dreams compared to awake has been variously characterized as bizarre, hallucinatory, or delusional. Researchers focused on dreaming as "hallucinatory" thought, may consider sleep mentation which resembles awake thought as non-dreaming. Sleep medicine generally considers the dream as any mentation reported as occurring during sleep, varying from the vague awareness of dreaming typical of the arousal disorders to the bizarre content of the classic psychoanalytic dream. The content axis can be viewed as an entwined continuum based on increasing complexity of associations as well as increasing variation from waking thought into the illogical, impossible, hallucinatory, bizarre, and delusional narratives which characterize some dreams.

If we cannot agree on a definition for dreaming we might as well digress into a discussion of how a dream is like gazpacho – cold or hot, tomato or not, soup made in many original, tasty, and foul ways. In both dream and in gazpacho, perceptual components involve the special senses, are variously defined and prepared, and, when extraordinary, integrate all levels of the psyche. Perhaps your view of gazpacho differs from mine. But as my daughter says, "Good gazpacho is to die for" (Fig. 3.2).

For the purpose of this book, we will use a definition for dream that addresses all three of the definition axes. We will address dreaming as mental activity (any mental activity) reported as occurring out of sleep. Of course, not everyone will agree that what this book is addressing is really dreaming. We will discuss the

FIGURE 3.2 A dream by definition (picture of gazpacho).

bizarreness of dreams and how they are like and unlike hallucinations. But we will not restrict dreams to being hallucinations. We will not restrict dreams to REM sleep. We will ask for a dream report to know that dreams have occurred. This perspective will restrict us to examining dreams in humans. Only humans can give us that dream report. But the definition of dreams remains fluid. Sometimes an impactful dream is as indescribable as the rushing border of an eclipse. Sometimes the dream report is its own story, filtered thorough the cognitive disarray of the artist. When asked for his definition of dream, the director John Sayles has said, "In high school I felt that I was living in a dream" (Pagel et al., 2003). That definition of dream is a story in itself, fitting in some ways much closer to the definition that Descartes himself used for dream back in the 17th century. High School can be a period of consciousness to which the scientific method does not apply!

The Biological Substrate of Dream

A substrate is the foundation or base that a biologic system requires for its existence. In this book we will delve into what we know of the human central nervous sys tem (CNS) as it applies to dreaming. At the very base of our scientific understanding are the neurochemistry, neuroanatomy, and electrophysiology of dream. By necessity, this presentation will need to be understandable to those outside specific sub-specialties of study as well as to the non-scientist. In our age of increasing specialization, it is as unusual for neuropathologist to understand CNS electro-physiology as it is for a non-scientist to understand neurochemistry.

There are few scientists that study dreams. It is not an area of study that attracts funding, tenure, or status. The scientific understanding of dreams is limited with few forums available for publication or presentation of the studies that reach completion. Yet almost everyone dreams. For many of us they seem important, and we search for understanding. Dreams seem to be powerful and often irrational probes extending into the borders of mind. The scientists who study them are as prone to anecdotes and personal memoirs of their own dreams as is any dreamer who describes his or her intensely personal dream story to another. In this section we try to avoid such personal content and present what is known of the scientific basis of dreaming: the brain matter, the electrical waves, and the neural-chemicals involved in dreaming. By necessity, since it has been the focus of so much study, we will examine research into rapid eye movement (REM) sleep in these areas of basic science. Restricting this presentation to evidence rather than theoretical con-jecture will mean for the reader that these presentations are surprisingly short.

In the final section chapter, we will take the basic science on step further into the emerging science of neural complexity, examining how theories of emergent con-sciousness may apply to dream. Some believe that sufficiently complex artificial intelligence systems will develop characteristics of self-awareness and mind. Perhaps even dreaming.

CHAPTER 4

The Neuroanatomy of Dreaming

You never identify yourself with the shadow cast by your body, or with its reflection, or with the body you see in a dream or in your imagination. Therefore you should not identify yourself with this living body, either.
(Shankara, 788–820 AD, Viveka Chudamani (Vedic Scriptures))

Any discussion of dreaming and neuroanatomy starts once again with rapid eye movement (REM) sleep.

The basic question comes down to whether there is something peculiar about the dream-associated neurophysiologic state that we call REM sleep. Dreams are reported at high frequency (approximately 80%) when individuals are awakened from REMS. But similar dream recall frequency is obtained on awakening immediately after sleep onset (Foulkes, 1985). Some authors argue that REMS dream content differs from the dreams of non-REMS sleep. REMS dreams are postulated to be more "dream-like," more prone to bizarre, hallucinatory mentation, the kind of dream sometimes incorporated into creative process (Hobson, 1994). Evidence to support this hypothesis has not been easy to come by. Domhoff (2003) has addressed the question of sleep stage specific dream content with extensive dream collection and analysis. He uses computer analysis of written dream reports to assess dream content based on the Van de Castle system of coding for words and effects (Hall & Van de Castle, 1966). This approach attempts to avoid the many problems of transference and researcher bias involved in the interpretation of dream content. Domhoff's studies indicate that most differences between REMS and non-REMS dreams can be accounted for by length of the dream report. The longer the dream report, the more likely the dream content is considered bizarre or hallucinatory.

Historically, sleep onset dreams are at least as "dream-like" as those reported on awakening from REMS. Famous creative dreams have occurred at sleep onset, including Descartes' dreams that led to the scientific method and Kekule's dream of snakes that led to the discovery of the ring chemical structure benzene. Since REMS occurs in normal individuals at least 60 minutes after sleep onset, it is unlikely that these are REMS dreams. Surrealist artists such as Salvador Dali were known to paint from sleep onset dream images. Dali would fall asleep in his chair with a coin in his hand. With the onset of sleep his voluntary muscles relaxed. When the coin hit the floor, he would wake and paint the image. Anyone believing that sleep onset dreams are not "dream-like," hallucinatory, or bizarre is encouraged to visit a museum or peruse a catalog of Dali's paintings.

THE NEUROANATOMY OF REMS

There is little question that viewed neuroanatomically, REM sleep is an electro-physiologic phenomenon that originates in specific sites of the brain stem. These sites designated by imaginative neuroanatomists are called the locus coeruleus and dorsal raphe nucleus. A modulation area for REMS is higher up the brain stem in mesopontine and pontine neurons (Pace-Schott, 2003). REMS neural interconnections have been documented in the hypothalamus, a brain area active in the process of memory; in the basal forebrain, involved in motor activity, vision, and memory; and in the amygala, a brain area involved in the processing of emotions (Hobson et al., 2003). Theorists supporting REMS = dreaming hypothesis speak of the activation and synthesis associated with dreaming. They propose that a cognitive dream develops in association with the projection of the REMS state into higher brain centers of the cerebral cortex. Positron emission tomography (PET) scans measure cellular glucose metabolism, a finding that reflects cellular activity occurring in the central nervous system (CNS). PET neuroimaging studies have shown that during REMS there is a general pattern of higher cortex deactivation (lateral prefrontal areas) and limbic midline activation (Nofzinger et al., 1997). It has been proposed that the executive functions of the frontal cortex are turned off by this prefrontal deactivation (Pace-Schott, 2003). This frontal cortex deactivation during REMS has been used to explain such postulated dream phenomenology as highly developed virtual realities associated with a lack of self-reflective awareness, organization, and attention. In Freudian parlance, the neuroscientist demonstrating that the higher cortex is shut down during REMS has demonstrated the absence of the super-ego in dreaming (Fig. 4.1).

But once again we are back to REMS. Is dreaming REM sleep? Do these PET scan findings for REM sleep reflect what is happening in the CNS during dreaming? PET scans of non-REM sleep consistently show decreased activity in other areas of the CNS compared to CNS activity during waking, including prefrontal, thalamic, and parietal areas. Perhaps this is dreaming as well.

It is much easier to study dreaming if it is assumed to be REMS. REM sleep occurs in almost mammals. In the laboratory neuroscientists use mice, rat, rabbit, and cat models in which specific CNS lesions affecting REM sleep can be induced. Single cell neural electrodes can be implanted that are able to detect the firing of single neurons. Because these approaches are widely available, neuroanatomists have been able to map the neural triggers and interconnections involved in REMS. It is much more difficult to study the neuropathology of dreaming when documentation of dreaming requires a verbal or written report of the dream. Then your studies are restricted to the single species that has the capacity to report whether are not dreaming has occurred. While researchers may be allowed to induce lesions in animals, they are only rarely allowed to use such approaches in humans.

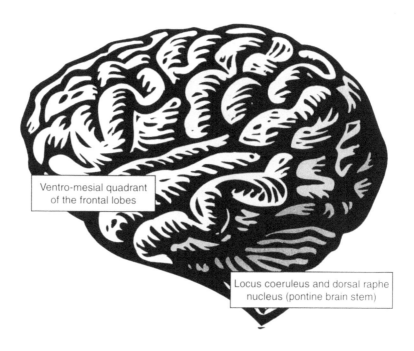

FIGURE 4.1 Damage to the ventro-mesial quadrant of the frontal lobes of the cerebral cortex can lead to a lack of dreaming. Damage to the locus coeruleus and dorsal raphe nucleus of the pontine brain stem can eliminate REMS in animal models but has not been shown to eliminate dreaming in humans.

New and improved brain scanning studies are available that can be used to study brain activity in humans. PET scanning can be done "real time" to detect what is going on in the brain when an individual attempts cognitive puzzles or expresses emotion. But even PET scanning cannot be used to detect dreaming. We still define sleep as a reversible state of perceptual isolation. The sleeping individual cannot interact with the researcher. Scanning techniques are unable to detect whether an individual is actually dreaming, because the individual cannot report dreaming when he or she is asleep. PET scanning studies with lucid dreamers – individuals that report an awareness of dreaming while dreaming – potentially offer the opportunity to collect CNS scanning data on dreaming individuals. However, it remains unclear whether individuals actually exist who can report dreaming during sleep by pushing a button or otherwise contacting the observer without waking up to do so. One of the characteristics of REM sleep is voluntary motor paralysis. Individuals able to push buttons are awake, according to both behavioral and electrophysiologic definitions.

THE NEUROPATHOLOGY OF DREAMING

The presence or absence of dream recall has been studied in human patients with CNS damage (Solms, 1997; Solms & Turnbull, 2002). REM sleep is a brain stem phenomenon. In cats brain stem lesions can be induced that eliminate REM sleep.

In humans, it is unusual for patients with extensive brain stem damage to survive that trauma or be able to interact to describe whether they still dream. However, in Solms' case series of more than 40 patients with extensive pontine brain stem damage, those patients still able to interact with the researcher continued to report dreaming. Individuals will continue to dream even after extreme levels of CNS damage. Individuals still report dreaming after having 1/2 of the cortex of their brain removed because of intractable seizures (Nielsen, 2003). There are, however, CNS lesions that do affect an individual's capacity to recall dreams. The large lesions that affect dreaming are those that affect the base area of the frontal cortex at the parieto-temporo-occipital junction area or in the ventro-mesial quadrant of the frontal lobes. These are the same lesions surgically induced in the 1940s and 1950s in a neurosurgical procedure called the modified prefrontal leucotomy. At one time such psychosurgery was a commonly utilized approach for the treatment of intractable psychiatric illness. Most of these patients (70–90%) reported complete or nearly complete loss of dreaming after surgery (Solms, 1997). Polysomnography (PSG) studies of these patients indicate that REM sleep is still present. These findings suggest that dreaming is a process requiring the frontal cortex.

The modified prefrontal leucotomy leads to other behavioral changes beyond the loss of dreaming. Emotional expression, drive, and behavior are often altered. Psychotic symptoms, when present, often improve with a decline in delusions and hallucinations (Benson, 1994). A typical spectrum of adverse effects associated with these psychosurgeries includes inertia, apathy, intellectual decline, personality change, and post-operative epilepsy (Solms and Turnbull, 2002). For any individual the behavioral entity called *personality* represents a stable state that is immensely difficult to alter through education, behavioral modification, or even psychotherapy. Frontal leucotomy is one of the few procedures known to consistently alter personality (Benson, 1994).

Some disease states are associated with a decline in dream recall. Mood disorders such as depression may profoundly decrease dreaming. Dream recall sometimes increases after treatment with antidepressant drugs (Armitage, 1994). The content of dreams in depressed individuals reflects the negative cognition of their waking state. Reduced dream recall is reported by individuals with obstructive sleep apnea as well. The basis for this change in dream recall may be secondary to diffuse CNS damage secondary to the effects of episodic hypoxia on the brain. Sleep apnea also causes severe daytime sleepiness. Sleepy individuals are less likely to recall dreams (Myers & Pagel, 2001). Treatment of sleep apnea may result in a return of dream recall (Pagel & Vann, 1997). Both depression and obstructive sleep apnea are very common disorders with global rather than specific effects on CNS function.

THE NEUROANATOMY OF DREAMING

So what do we know about the neuroanatomy of dreaming? If REMS is dreaming, we know a lot. REMS is triggered every 90 minutes or so during sleep by PGO waves affecting norepinephrine neurons in the locus coeruleus and dorsal raphe nucleus of the pontine brain stem. These neurons trigger the hippocampal theta frequency typical of REMS and are modulated by serotonergic mesopontine and pontine neurons. The cognitive characteristics of dreaming could be based on prefrontal cortex deactivation (a loss of executive cortical input resultant loss of self-reflective awareness, organization, and attention) and limbic (emotional) midline activation. These findings fit well with psychoanalytic models of the psyche, with REMS dreams uncontrolled by super-ego modulation, rising out of the primitive portions of the brain – the psychoanalytic id of the brain stem.

It would be easier for both the neuroanatomist and the Freudian psychoanalyst if REMS were dreaming, but that is not what the experimental evidence indicates. Dreaming occurs without REMS and REMS without dreaming. Dreaming is not restricted to REMS. We dream throughout the behavioral state of sleep.

According to PET scan results, both REMS and non-REMS sleep are associated with reduced blood flow in the prefrontal cortex. This turning off of what many deem to be the highest levels of cortical function may be associated with dreaming. But it would not be surprising if this prefrontal deactivation is characteristic of the state of sleep. Differences in scanning areas of activation noted between REMS and non-REMS may not reflect differences in dreaming. There are many other differences between REMS and non-REMS. REMS is an unusual state of sleep with unique physiologic characteristics. These include repetitive conjugate eye movements, penile and clitoral tumescence, respiratory and cardiac irregularity, and skeletal muscle paralysis. REMS is a state closer to waking than most of the other stages of sleep. It is common that an episode of REMS results in arousal from sleep. The scan derived patterns of CNS activation or inactivation associated with REMS are just as likely to result from these other REMS associated phenomena as they are from dreaming. All of these physiologic correlates of REMS generally occur in association with the state, but none are always associated with the state. Even the rapid eye movements that define the state can disappear for minutes of time. The classic correlate of penile erection occurs outside of REMS at least 50% of the time (Wasserman et al., 1982).

Sleep and electroencephalogram (EEG) electrical activity are universal experiences evidently required for healthy life and function. REMS remains a unique and unusual state of CNS activation and/or inactivation during sleep. Individuals can, however, function with little or no REM sleep. In the sleep laboratory, this is a common finding in patients with sleep apnea and those taking some psychoactive medications such as barbiturates and antidepressants. Yet REMS has been proposed to have more functions than the sleep state itself.

REMS has been evolutionarily preserved, and is present in almost all mammals except for the egg-laying monotremes of Australia in which it is difficult to document even the presence of sleep (Winson, 1972). A variant of REMS can be found in reptiles and birds. Humans REM sleep occurs in cycles every 90 minutes during the night, and it is often associated with recurrent episodes of waking. Visualize an early human sleeping under a tree with his mates on the plains of Africa, surrounded by large and hungry carnivores that hunt at night. In such a setting, these recurrent episodic awakenings associated with nocturnal cycles of REMS have been proposed to have survival value for the species (Vertes, 1986).

Roffwarg along with his cohorts (1966) have suggested that REM sleep functions in utero and infancy by approximating the neural interactions and connections required for conscious functioning before the infant is exposed to the external environment. Newborns spend up to 40% of their sleep in a variant of REMS called active sleep. This finding suggests that the REMS state is likely to have a role in development, perhaps as a rehearsal state for neural pathways that will later support perceptual input or perhaps as a template for instinctual behaviors (Roffwarg, 1966).

In adulthood, in our current world no longer requiring hunting and gathering on the plains of Africa, many studies have suggested that REMS has a functional role in the incorporation of waking experience into memory (Pagel et al., 1973; Smith, 1995). Mouse strains that as a group perform better on maze tasks have higher amounts of REMS than strains that perform poorly (Pagel et al., 1973). REMS appears to be associated with post-learning augmentation, the incorporation of experience into memory. Depriving an animal of REMS can disrupt prior learning (Smith, 1985). REMS appears to have an important cognitive role in our orientation response to novel experience (Morrison & Reiner, 1985).

REMS may have had an evolutionary function in inducing nocturnal arousals. Intrauterine, infant and childhood REM sleep is likely to function in infant neural development. Later in adulthood REMS functions in learning and memory. As the association of REMS with dreaming has become increasingly problematic, research and theory has returned to focusing on this important role for REMS. The association of REMS with learning and memory was apparent even in early studies of the state. Thirty-seven years after Dewan first enunciated his programming hypothesis for REMS, many cognitive scientists have turned their focus back to the role that REMS serves in the incorporation of experience into memory (Dewan, 1970).

REMS is a fascinating state and neuropathology and neuroanatomical research have taught us much about it. PGO waves affect norepinephrine neurons in the locus coeruleus and dorsal raphe nucleus of the pontine brain stem. These neurons trigger the hippocampal theta frequency typical of REMS and are modulated by serotonergic mesopontine and pontine neurons. REMS originates in the brain stem and has multiple cortical interactions inducing a spectrum of physiological effects that include the frequent report of dreaming on awakening from the state. But what this science has taught us is about REM sleep. It is not about dreaming. There is no simple causal relationship between REMS and dreaming.

We know some things about the neuroanatomy of dreaming. Dreams are preserved in individuals with extensive cortical damage. This finding suggests that dreams may have an important role in CNS functioning. Extensive damage to the base area of the frontal cortex at the parieto-temporo-occipital junction area or in the ventro-mesial quadrant of the frontal lobes can lead to a lack of dreaming. The damage that affects these areas destroys much of the communicative axons between right and left brain hemispheres. Individuals with damage to these areas have a spectrum of behavioral changes that extend beyond a lack of dream recall including personality change. It remains a question as to whether such damage truly leads to an elimination of dreaming or rather to an inability to recall or express whether dreams have occurred. We do know that the specific sites associated with a loss of dream recall are not those brain stem sites associated with a loss of REM sleep (Fig. 4.1).

CHAPTER 5

The Neurochemistry of Dreaming

Sleep is supposed to be,
By souls of sanity,
The shutting of an eye.

(Emily Dickinson)

The interactions between nerve cells are modulated by neural chemicals. These compounds are released at the synapse, the interface between neurons, and affect the tendency of the neighboring neuron to fire, carrying a neural impulse up the chain of dendrites and axons to the next cell in the neural network (Fig. 5.1). These compounds are called neurotransmitters. There are as many as 80 of these compounds. Recent data suggests that each neuron can utilize a spectrum of these neurotransmitters, some activating and some inhibiting, to communicate with neighboring neurons. The human central nervous system (CNS) is comprised of a hundred billion neurons, each with multiple synaptic connections and each with the capacity to respond to multiple neurotransmitters (Kandle, 2000; Swartz, 2000). This complex of neuroanatomy and neurochemistry is used to explain neuron functioning in cognitive processes.

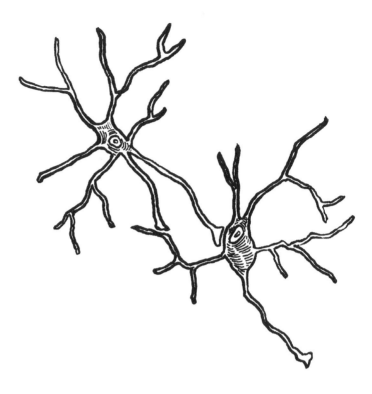

FIGURE 5.1 The Neural Synapse.

Just a few years ago, CNS active agents were classified as to their effects on behavior based on general cognitive effects of depression or excitation. Today, we know that the effects of most drugs on sleep and dreaming occur secondary to selective neurotransmitter effects. The neuronal systems modulating waking and sleep are contained within the isodendritic core of the brain extending from the medulla through the brain stem and hypothalamus up to the basal forebrain. Multiple factors and systems are involved, with no single chemical neurotransmitter identified as necessary or sufficient for modulating sleep and wakefulness. The fast-acting neurotransmitters of the CNS have many site-specific roles that may result in net changes on behavioral states and cortical arousal that directly contrast with their effects at the microscopic level. Today, almost all drugs with sleep and dreaming effects or sedative side effects can be shown to affect one or more of the widely dispersed central neurotransmitters important in the neuromodulation of sleep and wakefulness (Pagel, 2006).

Neurochemistry is a field structured on drug terminology that is often difficult and opaque for the non-chemist. Drug names are built from the first discovered compound in a class. New compounds may be denoted alphabetically (g-proteins, substance P) with the denotation marking the most chemically active of an experimental series of agents. The effort required to determine the actual structure of a compound is reserved for those agents such as the neurotransmitters proven to have physiologic effects. Research has focused on the primary

TABLE 5.1 Sedative–Hypnotics

Class	Drug	Sleep Stage Effects	Significant Side Effects
Benzodiazepines			
Short onset short half life – <4 hour	Triazolam (Halcion)	Decreased amplitude stages 3 and 4 Increased stage 2 (all) Shortened sleep latency In night REMS rebound	Loss of effect with chronic use Dependence antegrade amnesia
Short onset medium half life – 8.5 hours	Estazolam (ProSom)	Shortened sleep latency Decreased REMS	Daytime sleepiness
Short onset long half life – 50–110 hour	Flurazepam (Dalmane)	Shortened sleep latency Decreased REMS Withdrawal REMS rebound	Daytime sleepiness, chronic buildup (car accidents hip fractures)
Medium onset medium half life – 7–10 hour	Temezepam (Restoril) Clonazepam* (Klonopin)	Decreased REMS	Daytime sleepiness, poor sleep induction
GABA receptor agents			
Short onset medium half life	Zolpidem (Ambien) Eszopiclone (Lunesta)	Shortened sleep latency, benzo effects with dose above that normally perscribed	Idiosyncratic daytime sleepiness or antegrade amnesia
Short onset short half life	Zaleplon (Sonata) Indaplon		
Other agents			
Chloral hydrate	Chloral hydrate	Short sleep latency, decreased REMS, withdrawal REMS rebound	Low lethal dose, loss of effect with chronic use
Barbiturates and barbiturate like agents	Phenobarb, etc., methaqualone, glutethimide, ethechlorovynol	REMS suppression, short sleep latency, decreased REMS, withdrawal REMS rebound	Addiction low lethal dose, loss of effect with chronic use
	Methylprylon	Decreased sleep latency in some patients	Daytime sedation, anticholinergic
Sedating antihistamines H1-blockers	Diphenhydramine		
Melatonin agonists	Ramelteon	Shortened sleep latency	Neurohormonal interactions

neurotransmitters including dopamine, epinephrine and norepinephrine, acetylcholine, serotonin, histamine, glutamate, gamma-aminobutyrate (GABA), orexin, and adenosine. Many other chemical agents affect cognitive functioning in the CNS (Tables 5.1, 5.2, 5.3 & 5.4). The effects and/or side effects of these agents can directly affect the synapse or impact upon neurotransmitters active at those sites.

There are other neurochemical mechanisms through which drugs can affect neuronal function. However, these systems and compounds are generally not as well known or well described as the neurotransmitter systems. Some chemical agents exert their effects at the cell membrane, altering cellular membrane permeability and transduction, cellular energy through ATP production, or ionic flux.

Among these agents known to affect sleep and wakefulness are substance P, g-proteins, corticotrophin releasing factor, thyrotrophin releasing factor, nitric oxide, vasoactive intestinal peptide, melatonin, and neurotensin (Jones, 1998; Pace-Schott, 2003). Neuro-endocrine agents such as melatonin can also affect sleep and wakefulness, exerting effects outside the neural-transmission network.

REMS NEUROCHEMISTRY

Neurochemists interested in dreaming have concentrated on the effects of neurochemicals on rapid eye movement (REM) sleep. It might be expected from a state evolutionarily present in many species that the neurochemistry of REMS would be fairly simple. The original and quite simple model of REMS neurochemistry is called the reciprocal interaction model (Fig. 5.2) (McCarley & Hobson, 1975). This theoretical model describes the interplay between two major neurotransmitter systems (aminergic and cholinergic) involved in REMS generation in the brain stem. Subsequent work by the original authors and others has led to revised versions that incorporate the effects of other neurotransmitter systems that have been shown to affect the generation of REMS in the brain stem (Hobson et al., 2003). The authors' most recent version of this once simple system has become increasingly complex as other neurotransmitters and neuromodulators have been shown to affect REMS generation. The systems known to affect the generation of REMS include GABA, nitric oxide, glutamate, glycine, histamine, adenosine, dopamine, and other less well-described neuropeptides.

MEDICATIONS AFFECTING SLEEP

Many medications affect sleep. Medications that are used clinically to induce sleep (sedative–hypnotics) are some of the most widely prescribed medications available

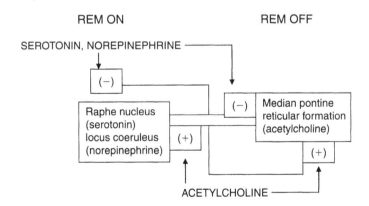

FIGURE 5.2 Original reciprocal interaction model.

in the world today. Most of these medications work by lowering levels of CNS arousal and allowing an individual to initiate sleep. These drugs have specific neurochemical and electrophysiological effects (Table 5.1). Antidepressants are often used to treat disturbances in sleep and mood. These medications can induce both sleepiness and insomnia based on specific neurotransmitter effects (Table 5.2). Drug-induced sleepiness is probably the most commonly reported side effect of a wide variety of other medications (Table 5.3). Drug-induced insomnia is reported somewhat less commonly as a side effect to medications, however, multiple groups of medications can cause insomnia in some patients (Table 5.4). Many of the drugs reported to induce sleepiness and/or insomnia as effects or side effects are the same agents reported to induce nightmares and disordered dreaming (Table 5.5) (Pagel & Helfer, 2003; Pagel, 2006b).

TABLE 5.2 Antidepressants (Sedating Agents in Bold)

Class	Drug	Sleep Stage Effects	Indications
Tricyclic	**Trimipramine, nortriptyline, doxepin, amoxapine, amitryptyline, imipramine, amoxapine, protriptyline***	Increased – REMS latency Decreased – REMS(++), SWS latency Deep sleep, sleep latency	Depression with insomnia, REMS and SWS supression, chronic pain, fibromyalgia, enuresis (etc.)
Non-tricyclic sedating	**Desimprinine, maprotiline, mirtazapine**	Increased – REMS latency Decreased – SWS latency, REMS(++), sleep latency	Depression, depression with insomnia, REMS suppression
MAOI	**Phenelzine, tranylcypromine**	Increased – stage 4 Decreased – REMS latency, REMS (+++)	Depression, REMS suppression
SSRI	*Fluoxetine**, paroxetine, *sertraline*, fluvoxamine, citalopram HBR	Increased – REMS latency, sleep latency, stage 1 Decreased – REMS	Depression, PTSD, obsessive compulsive disorder, phobias, cataplexy (etc.)
Serotonin + Norepinephrine reuptake inhibitor	*Venlafaxine*	Increased – REMS latency Decreased – sleep latency, REMS	Depression
Dopamine + Norepinephrine reuptake inhibitor	Bupropion	Increased – REMS latency, sleep latency	Depression, nicotine withdrawal
Non-tricyclic non-SSRI	**Nefazodone**	Increased – REMS Decreased – sleep latency	Depression, depression with insomnia and anxiety
Serotonin 1A agonist	Buspirone	Increased – REMS latency Decreased – REMS	Anxiety

Key: (++) Higer levels of effect; (*) documented as respiratory stimulant.

TABLE 5.3 Medication Types Reported in Clinical Trials and Case Reports to Have Sleepiness as a Side Effect

Medication Class	Neurochemical Basis for Sleepiness
Antihistamines	Histamine receptor blockade
Antiparkinsonian agents	Dopamine receptor agonists
Antimuscarinic/antispasmotic	Varied effects
Skeletal muscle relaxants	Varied effects
Alpha-adrenaergic blocking agents	Alpha-1-adrenergic antagonists
Beta-adrenergic blocking agents	Beta-adrenergic antagonists
Opiate agonists	Opioid receptor agonists (general CNS depression)
Opiate partial agonists	Opioid receptor agonists (general CNS depression)
Anticonvulsants	
Barbiturates	GABA receptor agonists
Benzodiazepines	GABA receptor agonists
Hydantoins	General effects?
Succinimides	General effects?
Other	Varied effects including GABA potentiation
Antidepressants	
MAOI	Norepinephrine, 5HT, and dopamine effects
Tricyclic	Acetylcholine blockade, norepinephrine, and 5HT uptake inhibition
SSRI	5HT uptake inhibition
Others	5HT, dopanine, and norepinephrine effects
Antipsychotics	Dopamine receptor blockade, varied effects on histaminic, cholinergic, and alpha-adrenergic receptors
Barbiturates	GABA agonists
Benzodiazepines	GABA agonists
Anxiolytics, miscellaneous sedative and hypnotics	GABA agonists, varied effects
Antitussives	General?
Antidiarrhea agents	Opioid, general?
Antiemetics	Antihistamine and varied effects
Genitourinary smooth muscle relaxants	General?

TABLE 5.4 Medication Types Known to Cause Insomnia

Adrenocorticotropin (ACTH) and cortisone
Antibiotics – Quinolones
Anticonvulsants
Antihypertensives (alpha-agonists, beta-blockers, central-acting agents)
Antidepressants (SSRIs)
Antineoplastic agents
Appetite suppressants
Beta-agonists
Caffine
Decogestants
Diuretics
Dopamine agonists
Ephedrine and psuedoephedrine
Ethanol
Ginsing
Lipid and cholesterol lowering agents
Niacin
Oral contraceptives
Psycho-stimulants and amphetamines
Sedative/hypnotics
Theophylline
Thyroid preparations

Affected Neuroreceptor Drug	Patient Reports of Nightmares – Evidence Base Clinical Trials (CT), Case Reports (CR)	Probability Assessment of Drug Effect
Acetylcholine–cholinergic agonists		
Donepezil	CT (3/747 report disordered dreaming)	Possible
Norepinephrine – beta-blockers		
Atenolol	CT (3/20 patients)	Probable
Bisoprolol	CT (3/68 patients): CR [1] – de-challenge	Probable
Labetalol	CT (5/175 patients)	Probable
Oxprenolol	CT (11/130 patients)	Probable
Propranolol	CT (8/107 patients)	Probable
Norepinephrine effecting agents		
Guanethidine	CT (4/48 patients)	Probable
Serotonin – SSRI		
Fluoxetine	CT (1–5% – greater frequency in OCD and bulimic trials: CR (4) – de- and re-challenge	Probable
Escitalopram oxylate	CT (abnormal dreaming – 1% of 999 patients)	Probable
Nefazodone	CT (3% (372) versus 2% control)	Probable
Paroxetine	CT (4% (392) versus 1% control)	Significant
Agents affecting serotonin and norepinephrine		
Risperidone	CT (1% increased dream activity – 2607 patients)	Probable
Venlafaxine	CT (4% (1033) versus 3% control)	Probable
Dopamine – agonists		
Amantadine	CT (5% report abnormal dreams): CR (1)	Probable
Levodopa	CT (2/9 patients)	Probable
Ropinirole	CT (3% (208) report abnormal dreaming versus 2% placebo)	Probable
Selegiline	CT (2/49 reporting vivid dreams)	Probable
Amphetamine like agents		
Bethanidine	CT (2/44 patients)	Probable
Fenfluramine	CT (7/28 patients): CR (1) de- and re-challenge	Probable
Phenmetrazine	CT (3/81 patients)	Probable
GABA		
GABA hydroxyl buterate	CT (nightmares >1% 473 patients)	Probable
Triazolam	CT (7/21 patients)	Probable
Zopiclone	CT (3 to 5/83 patients)	Probable
Antiinfectives and immunosupressants		
Amantadine	CT (5% reporting abnormal dreams): CR (1)	Probable
Fleroxacin	CT (7/84 patients)	Probable
Ganciclovir	CR (1) – de- and re-challenge	Probable
Gusperimus	CT (13/36 patient)	Probable
Antipsychotics		
Clozapine	CT (4%)	Probable
Antihistamine		
Chlorpheniramine	CT (4/80 patients)	Probable
Ace inhibitors		
Enalapril	CT (0.5–1% abnormal dreaming – 2987 patients)	Probable
Lovistatin	CT (>1% dream abnormality – 858 patients)	Probable
Quinapril	CT	Probable
Other agents		
Digoxin	CR (1) – de- and re-challenge	Probable
Naproxen	CR (1) – de- and re-challenge	Probable
Verapamil	CR (1) – de- and re-challenge	Probable

MEDICATIONS AFFECTING DREAMING

Few if any studies have looked at the effects of medications on dreaming in human beings, the only species that can report both the content and an experience of whether a dream has occurred. The medications proposed to affect dreaming are the same ones known to affect REM sleep effects as based on the reciprocal interaction model as outlined above. Agents that suppress REMS such as ethanol and benzodiazepines such as valium induce episodes of REMS rebound during medication withdrawal. These REMS rebound episodes have been associated with reports of nightmares and disturbed dreaming, and have generally been considered the primary mechanism resulting drug-induced disordered dreaming and nightmares.

In the last 15 years, however, clinical trials required for the marketing of new pharmacological agents have asked the patients involved in these studies to report nightmares and alterations in dreaming. Physicians in medical practice can also submit case reports of drug side effects for drugs in clinical use. These case reports sometimes include the complaint of nightmares and/or disordered dreaming. This information can be coupled with clinical trial data of medications noted to induce nightmares. The primary limitations of this approach is that it does not include older drugs for which nightmares were not monitored in clinical trials or other agents that are not approved for clinical use in the practice of medicine. Based on these patient reports of disturbed dreaming rather than theory, a pattern of medication classes noted to induce nightmares becomes apparent. This data suggests that the medications associated with clinical reports of disordered dreaming differ from those postulated to induce nightmares based on the association of dreaming with REM sleep rebound. The neurochemistry of dreaming appears to be more complex than even the modified reciprocal interaction model of REM sleep (Hobson et al., 2003; Pagel & Helfer, 2003; Pagel, 2006b).

PRIMARY NEUROTRANSMITTERS AFFECTING SLEEP AND DREAMING

Acetylcholine

The electrical nature of the action potential conveying nerve impulses from neuron to neuron was first described by Ramon y Cajal and Charles Sheridan at the turn of the last century. The chemical mediator of that spike potential between neurons was not clarified until the 1930s. The Austrian pharmacologist Otto Loewi had postulated the possibility that the effects of the vagal nerve on heart rate were chemically mediated. He could not, however, determine how to experimentally prove his postulate. One night he awakened after a dream in which he was sure that he had discovered the experimental solution to the problem. But try as he might, he was unable to remember his dream. The next night he went to bed intent on re-dreaming the solution. He woke and rushed to the lab where he electrically stimulated the vagus nerve of a frog to induce a slowing of the heart rate. He took the blood from that frog and injected it into another, inducing in that frog a slowing of the heart rate as well. This demonstrated that the slowing of the heart rate caused by stimulation of the vagus nerve was mediated by a chemical in the blood. That chemical, acetylcholine, became the first neurotransmitter to be isolated.

REM sleep is affected by pharmacological alteration of acetylcholine activity in the CNS. There are several lines of evidence supporting the conclusion that brain stem cholinergic neurons can be excited to induce REM sleep (Steriade, 2000). Cholinergic agents are most likely to increase percentages of REM sleep, with cholinergic antagonists tending to decrease REM sleep (Hobson & Steriade, 1986). Anticholinergics are among the neurochemical agents that suppress REM sleep in humans. A wide variety of pharmaceutical agents have anticholinergic activity. The reported side effects of some of these agents include nightmares, disordered dreaming, and hallucinations. This has led some authors to postulate that it is the cholinergic effects of medications that lead to psychiatric side effects such as hallucinations or psychosis (Perry & Perry, 1995). Recently, several agents with anti-cholinesterase effects have come into widespread use for the treatment of the cognitive effects from early Alzheimer's disease. These agents increase acetylcholine in the CNS by blocking anti-cholinesterase, the primary system utilized in the breakdown of acetylcholine. Despite the known role of acetylcholine in the initiation of REM sleep, these medications known to affect acetylcholine are rarely reported to affect dreaming. The side effect of disturbed dreaming or nightmares was reported by only 3 of 747 patients taking the most commonly used of these anti-cholinesterase drugs – donepezil (Aricept) – in clinical trials (Table 5.2) (Pagel & Helfer, 2003).

Norepinephrine

Many of the drugs in general use for treating high blood pressure (hypertension) affect norepinephrine receptors. These drugs have been shown to affect both REM sleep and reports of dreaming. Because these agents suppress REM sleep, they are sometimes used clinically in the treatment of recurrent nightmares in patients with post-traumatic stress disorder (PTSD). Yet the norepinephrine affecting antihypertensive agents classified as beta-blockers and alpha-agonist are responsible for 34% of clinical trials in which nightmares are reported as an adverse effect (Thompson & Pierce, 1999). The reported effects of these agents on both dreams and nightmares are often opposite to the drug's known pharmacological effects on REM sleep. Decreases in dream recall occur with use of both alpha-agonists (ex. minoxidil (Minipress)) that are REM suppressant and beta-blockers (ex. propranolol (Inderal), and atenolol (Tenormin)) that do not suppress REM sleep. The use of beta-blockers depresses REM sleep percentages yet can result in reports of increased dreaming, nightmares, and hallucinations (Dimsdale & Newton, 1991). The effects of these agents demonstrate that a drugs' effect on REM sleep may or may not be associated with an associated change in reported dreaming.

Serotonin

Both serotonin and norepinephrine have functional roles in the production of REM sleep (Fig. 5.2). The neurochemical action of many antidepressants is primarily to increase serotonin levels. However, some agents affect a variety of other neurotransmitters (Table 5.3). Most antidepressants suppress REM sleep. This effect is greatest for the older types of antidepressants including the monoamine oxidase inhibitors (MAOIs) and the tricyclic antidepressants (ex. Amitryptyline (Elavil) and imipramine (Tofranil)). However, even the newer selective serotonin reuptake inhibitors (SSRIs) (ex. paroxetine (Paxil) and sergeline (Zoloft)) are potent suppressors of REMS). REMS suppression is not generally seen with buspirone (Wellbutin). Most antidepressants are reported in clinical trials to induce nightmares in some patients. Case reports of nightmares are associated with fluoxetine (Prozac) (Gursky & Krahn, 2000). Intense visual dreaming and nightmares are associated with the acute withdrawal from some antidepressants (Coupland et al., 1996). This effect could be due to REM sleep rebound occurring after the withdrawal of these REM sleep suppressant agents. However, studies of reported dream recall with antidepressant use show that recall may vary independently of REM sleep suppression (Pace-Schott et al., 1999). Studies of chronic steady state use and antidepressant withdrawal have shown inconsistent effects with increased dream recall, no effect, and decreased recall reported in different studies utilizing SSRIs and tricyclics (Lepkifkier et al., 1995).

Dopamine

Dopamine receptor stimulation is another common mechanism resulting in drug-induced nightmares. Dopaminergic medications are commonly used in the treatment of Parkinson's disease and restless leg syndrome. These medications share the tendency to cause disordered dreaming. Dopamine (Sinnemet), bromocriptine (Permax), pergoline (Requip), and other dopamine agonists can lead to vivid dreaming, nightmares, and night terrors which can be the first signs of the development of drug-induced psychosis (Stacy, 1999). The amphetamines exert their effects at dopamine receptors. Amphetamine use has been linked to nightmares (16% of nightmare reports from clinical trials). This effect has been postulated to occur secondary to dopamine receptor stimulation (Thompson, 1999).

GABA

GABA is the primary negative feedback neurotransmitter in the CNS. Most hypnotics (sleep inducing) agents affect this receptor. Some neurochemists refer to the GABA receptor as the benzodiazepine (ex. diazepam (Valium)) receptor since this is the site where these agents exert their primary neurochemical effects. Twenty-four percent of the reports of nightmares come from benzodiazepine clinical trials (Thompson & Pierce, 1999). The newer non-benzodiazepine hypnotic (eszopiclone (Lunesta)) which is not associated at clinical dosages with REM sleep suppression or REM sleep rebound on withdrawal has been associated with the occurrence of nightmares in several clinical trials as have newer agents affecting GABA reuptake inhibition (Mitler, 2000).

NEUROTRANSMITTER MODULATING SYSTEMS

Neuronal populations utilizing the neurotransmitter acetylcholine, serotonin, and norepinephrine have prominent roles in the control of the REM–non-REM sleep cycle (Fig. 5.2) (Hobson et al., 2003). These neurotransmitters may function in a reciprocal interaction that also involves a wide spectrum of other neurotransmitters that interact in an intricate modulation of the stages of sleep (Pace-Schott, 2003). Proposed neurotransmitter modulators affecting this system include

GABA, dopamine, orexin, adenosine, histamine, glycine, glutamate, nitric oxide, and neuropeptides. In addition to antidepressants and benzodiazepines, clinical REM sleep suppressants include ethanol, barbiturates, and sympathomimetic drugs (Pace-Schott, 2003; Hobson et al., 2003). All of these agents have been shown to affect the modulation of this REMS ON/REMS OFF system in the brain stem and therefore affect REM sleep.

DREAMING AND NIGHTMARES: NEUROTRANSMITTER SYSTEMS

If we look at patient reports of disturbed dreaming, we find that some of these same neurotransmitters known to affect REMS sleep also affect dreaming. However, the effect of these agents is quite different than we would expect based on theory (Fig. 5.2). Almost all of the agents exerting their neurochemical effects on dopamine, serotonin, and norepinephrine will induce altered dreaming and nightmares in some patients. Among prescription medications in clinical use, beta-blockers affecting norepinephrine neuroreceptors are the agents most likely to result in patient complaints of nightmares. The strongest clinical evidence for a drug to induce disordered dreaming or nightmares is for the SSRI paroxetine (Paxil), a medication known to suppress REMS. Most agents affecting dopaminergic neuroreceptors induce nightmares in some patients. Medications altering these neurotransmitter systems are likely to induce reports of nightmares and disordered dreaming for patients taking those medications.

The association of GABA and acetylcholine receptors with dreaming and nightmare alteration is less clear. The finding that different types of drugs known to affect the GABA receptor (agonists, modulators, and reuptake inhibitors) can result in patient complaints of nightmares and abnormal dreaming suggests that GABA may be a modulator of the neuronal populations involved in dreaming (Mallick et al., 2001). Acetylcholinesterase inhibitors affecting the acetylcholine neuroreceptor system rarely result in patient complaints of drug-induced nightmares. Studies of drug effects and side effects do not provide good support for theoretical postulates that cholinergic neurons (the triggers for REMS) serve as the primary neuroreceptor system involved in dreaming and nightmares (Fig. 5.2) (Pagel & Helfer, 2003; Pagel, 2006b).

Other neurotransmitter modulators proposed to affect sleep and dreaming include orexin, adenosine, histamine, glycine, glutamate, nitric acid, and neuropeptides (Pace-Schott, 2003). The commonly used antihistamine chlorpheniramine (Tylenol PM, Benadryl) induces nightmares in some patients suggesting a potential role for histamine as a modulator of dreaming. The neurochemical and pharmacological basis for clinical effect for many of the agents included in Table 5.3 remains poorly defined. It is possible that the induction of nightmares and altered dreaming by some of these agents is secondary to neurotransmitter effects that have yet to be described.

AGENTS AFFECTING CONSCIOUS INTERACTION WITH THE ENVIRONMENT (ANESTHETICS)

Agents which alter an individual's conscious relationship to the external environment are known to alter dream and nightmare occurrence. Although not clearly sleep inducers, many of the agents reported to cause altered dreaming are induction anesthetics utilized in surgery. An increased incidence of "pleasant" dreams are reported with the use of profolol as an anesthetic in surgery (Marsh et al., 1992). The barbiturate thiopental, ketamine, and the opiate tramadol have produced disordered dreaming and nightmares (Krissel et al., 1994; Oxorn et al., 1991). Some of the agents associated with the complaint of nightmares also can induce waking hallucinations and confusion (fleroxacin, triazolam, ethanol withdrawal, and amphetamines) (Pagel & Helfer, 2003). This association has, in part, led to proposals that dreams and nightmares are hallucinatory experiences occurring during sleep (Hobson, 1999).

AGENTS AFFECTING HOST DEFENSE

Aristotle and Hippocrates first pointed out that an association exists between infection and sleepiness. Both viral and bacterial infections can be associated with severe somnolence and large increases in non-REM sleep (Krueger & Fang, 2000). Such microbial-induced changes in sleep are considered part of the acute phase response to infection. Both muramyl peptides and endotoxins are chemical mediators of infection that have been shown to induce sleepiness (Krueger et al., 1986). Some of the antibiotics (i.e., fluoroquinolones (Floxin and Cipro)) that are known to induce insomnia have also been reported to induce nightmares. Other chemical mediators involved in the inflammatory response to infection (the cytokines IL-1B and TNF-a, and prostaglandin E2) are known to be involved in non-REM sleep regulation (Jaffe, 2000). Infectious diseases are sometimes associated with the complaint of nightmares. Sleep loss affects host defense and cellular immune function. A diverse group of antibiotics, antivirals, and immunosuppressant drugs can induce the complaint of nightmares for some patients. Host defense interacts with infectious disease to induce cognitive effects on sleep and dreaming for these agents. A strong but currently poorly defined relationship exists between host defense and infectious disease and sleep and dreaming.

THE NEUROCHEMISTRY OF DREAMS AND NIGHTMARES

The group of pharmacological preparations, both psychotropic and otherwise, reported to alter dreaming and induce nightmares is extensive and quite diverse. Medications that alter the neurotransmitters norepinephrine, serotonin, and dopamine consistently induce the complaint of disturbed dreaming and nightmares in some patients taking the drugs. However, the effects of these agents on dreaming may be opposite to the known effects of these agents on REM sleep. Agents affecting the neurotransmitters GABA, acetylcholine, and histamine are less likely to be associated with the complaint of nightmares.

Some medications appear to alter nightmare reporting by affecting an individual's conscious relationship to the environment (anesthetics) or host defense and immunology. Many of the agents reported to induce nightmares also induce CNS side effects of daytime somnolence and insomnia. Nightmare induction by agents affecting histamine and inducing daytime sedation as a side effect suggests that these agents are also likely to affect dreaming. Most drugs reported to affect dreaming also affect sleep and waking consciousness. The tendency of drugs to induce such cognitive side effects may be an indicator for drugs likely to induce disordered dreaming and nightmares (Tables 5.3 and 5.5).

The fact that a wide spectrum of pharmacological agents are reported to induce disturbed dreaming and nightmares suggests that the biochemical basis for dreaming is more complex and less understood than is generally assumed. This data based on patient reported nightmares and disturbed dreaming does not support the theory that cholinergic triggers inducing dreaming based on the reciprocal (REMS ON)–(REMS OFF) system (Fig. 5.2). However, drugs affecting serotonin and norepinephrine are among those most likely to alter patient reports of dreaming. The effects of medications on dreaming or nightmares are often opposite to the known effects of these agents on REM sleep. This finding is more evidence that dreaming is not a simple or derivative state of REM sleep.

Dreaming is better visualized as a complex state of sleep-based cognition that is poorly described by our current models of neuroanatomy and neurochemistry.

Medications that have the clinical effects of arousal (insomnia) or sedation are the medications most likely to alter dreaming and induce nightmares. This finding may be due to the fact that sleep is the substrate for dreaming, and medications that alter sleep are also likely to alter cognitive activity occurring in sleep such as dreaming. The effects of sedative medications may also act by affecting an individual's ability to recall a dream. Interesting, sedating medications often lead to an increase in the recall of dreams and nightmares while illnesses that cause daytime sleepiness such as depression and sleep apnea suppress dream recall. It is possible that sedative medications affect chemical systems that normally suppress the recall of dreams resulting in a disinhibited cognitive state on arousal in which dreams are more likely to be remembered. It is also possible that these medications are more likely to induce negative dreams such as nightmares. Nightmares are powerful, significant dream experiences that often wake the dreamer. Nightmares are among the types of dreams most likely to be remembered after waking in the morning.

What then is the basis for the neurochemistry of dreaming? It is unclear whether the drugs that affect dreaming are altering dreaming itself or affecting the ability to recall dreams. Neurochemically, dreaming is as complex as waking consciousness. Based on its neurochemistry, dreaming appears to be another state of consciousness, variably assessable in waking, and affected by the same groups of medications that can alter our cognitive interaction with the world.

CHAPTER 6

The Electrophysiology of Dreaming

And if he were made to look directly into the light, would this not hurt his eyes, and would he not turn back and retreat to the things which he had the power to see, thinking that these [the shadows] were in fact clearer than the things now being shown to him?

(Plato's Republic – *2nd Stage of the Cave*, 515 e 5)

Throughout sleep, we dream. Behaviorally, sleep is a reversible dissociation from the exterior environment. This global state can be divided into different stages based on its electrophysiological characteristics. Sleep staging developed as an artificial construct, based on the technological systems available for recording physiologic changes occurring during sleep. Electrical data from the various telemetry channels of the polysomnogram are used to categorize the changing electrical patterns of sleep occurring through the night. This telemetry data is used to divide sleep into the different sleep stages of rapid eye movement (REM) sleep and stages one through four of non-REM sleep (Fig. 6.1).

Over the last 50 years, dreaming has come to be defined by the non-conscious electrophysiologic correlate of dreaming that is called REM sleep or rapid eye movement sleep based on the repetitive conjugate eye movements that typically occur during this state of sleep. The assumption that REM sleep is dreaming has allowed researchers to study "dreaming" without requiring a cognitive report of dreaming. Since most animals have REMS, such "dreaming" can be studied in animal models. Real-time brain radioactive scanning techniques used to describe specific sites of central nervous system (CNS) neural activity associated with REM sleep have been postulated to show the specific brain activity associated with dreaming. Reflecting the belief that REMS = dreaming, the neural activity found associated with REM sleep has been hypothesized to be neural activity of dreaming.

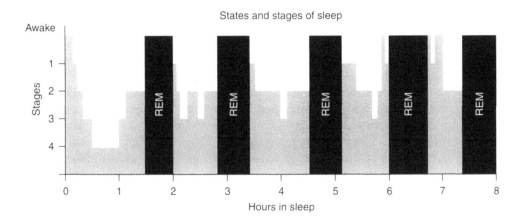

FIGURE 6.1 The electrophysiologically defined states and stages of sleep occurring across a night of recording.

However, we know that cognitive dreaming occurs in sleep stages that are not REMS. We also know that REMS occurs without dreaming. REM sleep is but one of the sleep states associated with a high frequency of dream recall on waking. REM sleep is better characterized as one among many electrophysiological markers commonly associated with dreaming (Domhoff, 2003; Foulkes, 1985). This fact that REMS and dreaming are doubly dissociable calls into question any experimental work and theory that is based on the supposition that REMS is equivalent to dreaming.

Recent research has begin to focus on CNS and behavioral correlates of dreaming other than REM sleep. High dream recall frequency is also reported at sleep onset, and with impending arousal from sleep. Dreams occur in association with other electrophysiological events during sleep. These electrophysiological markers also associated with REMS have not been generally used to define the state because they are more difficult to record in human subjects. These markers include pontine electrical burst activity called PGO spikes, hippocampal theta EEG (electroencephalogram) frequencies recorded using intracerebral EEG monitoring, and saw-tooth waves of the EEG that occur during REMS in some subjects. Dreaming is also associated with slow as well as with fast ocular movements, and with penile erection in males and clitoral erection in females. These physiologic correlates can be monitored and electrically recorded.

THE EEG

The development of the electrical amplifier in the 1920s led to the ability to record the brain electrical activity (in μV) measurable through electrodes attached to the human scalp. This monitoring technique provides the patterns of electrical activity that are reported as an EEG. Brain electrophysiological events are based on these extracerebral potential changes that we can record. Using a string galvanometer in 1928, Hans Berger first recorded the electrical activity of the alpha frequency rhythm associated with drowsiness and sleep onset. Today, EEG analysis is used in the clinical setting to assess the presence or absence of seizure disorders, for polysomnographic sleep staging, and to determine the presence or absence of CNS neural activity after cerebral injury. The origin of these electrical potentials that are recorded on the EEG remains a topic of debate in the field of brain research.

Neurochemistry and neuropathology have been extremely useful techniques for defining and clarifying neural interactions. The therapeutic approaches based on these neurosciences form the basis for the medical fields of neurosurgery and neurology. These sciences are the basis for neurochemical and neurosurgical approaches to the treatment of disease. However, our understanding of CNS electrophysiology remains limited. The origin of EEG electrical activity, particularly the background EEG rhythms that dominate the EEG during drowsiness and sleep, remains unclear. These background EEG rhythms are the electrophysiologic markers used to determine when non-REM states (stages 1–4) occur during sleep. Many scientists accept that the electrical background frequencies and spikes seen on the EEG are an index of neural activity occurring in the brain. Changes in this EEG activity can signify disease such as epilepsy or coma. However, many of the same neuroscientists also believe that the background electrical activity described by the EEG has no inherent physiological function. Despite what many accept as a lack of function for the EEG rhythms, the absence of this activity is used as a marker for organism death. There is a level of cognitive dissonance in the realization that these non-functional EEG rhythms are legally and physiologically considered to be essential markers of life.

FUNCTIONAL ROLES FOR CNS ELECTRICAL FIELDS

Viewed as an electrophysiologic process, sleep can be divided into stages based on the occurrence of synchronous EEG wave activity. This is but one of the ways that sleep differs from waking. In waking such wave-like EEG activity is rare. Drowsy awake with eyes closed and stage 1 (sleep onset) are both electrophysiologically dominated by the presence of waves of alpha rhythms occurring at 8–12 cycles per second. This frequency is the dominant and most common of the brain rhythms. Based on the spectral analysis of sleep, alpha has by far the greatest power of the physiologic brain rhythms. Stage 2 sleep is denoted by bursts of sleep spindles at sigma frequency of 14–15 cycles per second. Deep sleep (stages 3 and 4) occurs in association with delta frequency oscillations of 0.5–1.5 cycles of second. REMS is described by some authors as "desynchronized" with less apparent waveform activity than the non-REM stages. However, REMS only appears to be desynchronized because of the methods that are used to record the human EEG. Human EEG monitoring uses high impedance skin electrodes on the face and scalp to record the low voltage physiologic EEG potentials (<50 mV) as well as the associated eye movements (electrooculogram, EOG) and electrical activity of muscles (electromyography, EMG) required to define the state. In animal models and in human subjects during neurosurgery, deep EEG monitoring sites can be utilized. REMS recorded intracranially is dominated by long runs of hippocampal waves of theta frequency occurring at 5–8 cycles per second. This is an extremely synchronous rhythm. Recorded in this manner, REMS has a clear association with CNS electrical wave activity and can be considered the most synchronous of sleep stages (Fig. 6.2) (Siegel, 2000).

Francis Crick of DNA research fame has suggested that the functional CNS rhythms responsible for conscious experience are not these physiologic rhythms normally recorded during sleep. He proposes that consciousness is associated with rhythms occurring at the higher frequency of 40 cycles per second (Hz) (Chalmbers, 1996; Crick & Mitchinson, 1983). This is a frequency that we do not normally see or record with our modern EEG equipment. In our modern electrically enervated world 60-cycle background electrically activity fills our environment with the alternating electrical currents of the electrical power grid. This electrical artifact hides mid/high range frequencies such as 40 Hz. In the sleep laboratory, electrical filters are used to block out the electrical activity occurring in this range so that the low intensity electrical activity of the brain can be recorded. It is possible that functional 40 Hz brain activity exists, but it is very difficult to record in our modern world. It can be argued that if hidden electrical fields function in this range, our constant exposure to powerful external electrical fields of equivalent frequencies might be expected affect our brain functioning. If our consciousness works using the same frequency range as our toasters and our

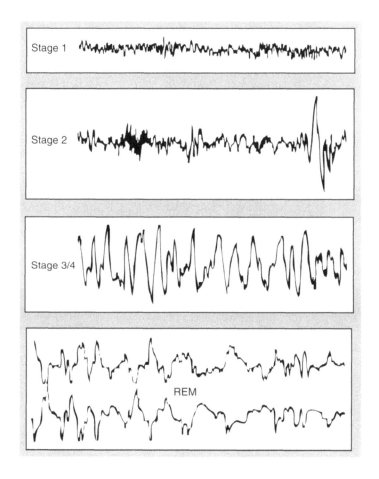

FIGURE 6.2 EEG electrical rhythms of the sleep stages.

electrical infrastructure, electrical interference is potentially responsible for a broad spectrum of modern neurological ills and discomforts.

It seems logical that the electrical potentials that we see on the EEG come from the neural electrical activity involved in CNS functioning. It is reasonable to propose that the EEG reflects the electrical activity occurring in the CNS, that the EEG is an extracellular summation of the individual electrical spike potentials utilized by neurons to interact. It has been difficult, however, to demonstrate how the discrete spike potentials of individual neurons firing can lead to propagated global electrical rhythms such as those recorded by the sleep EEG (Table 6.1) (Pagel, 2005).

Neuroscientists have yet to find an electrical rhythm generator in the CNS. Theoretically, specific physiologic frequencies could occur based on the time sequence of oscillatory opening and closing of ionic gateways and channels involved in the production of neural spike potentials. The time required for opening and closing chemical potassium-based gates could reflect the time sequence of neuron firing to produce a global alpha rhythm. The longer time required for opening and closing calcium-based ionic gateways could produce the sigma frequency (Cheek, 1989; Christakos, 1986; Steriade, 2001). This would mean that the physiological EEG frequencies are globally generated throughout the CNS as a result of the electrochemical neural membrane processes involved in neuron firing. The repetitive firing of neurons at set frequencies could potentially create self-generating feedback loops that result in a global tendency for neurons to fire in sequence.

The physiological systems that we can detect have functions. It would quite unusual if a physiologic system such as the brain electrical frequencies had no function. There are suggestions that these electrical frequencies affect nerve cells. It has been shown that changes in the physiological EEG rhythms can affect the tendency of an individual neuron to fire and produce a spike potential (John & Swartz, 1978). Hodgkin, Katz, and Goldman received the Nobel Prize in the 1950s for describing the way in which ion flux at the cell membrane can induce an electrical spike potential in a neuron. This effect is electrophysiologically described by the classic Hodgkin–Katz–Goldman equation for the interaction of potassium, sodium, and chloride ions involved in inducing a neural spike potential (Formula 6.1) (Hodgkin & Horowicz, 1959).

TABLE 6.1 Differences Between Physiologic EEG Rhythms and Spike Potentials

	Voltage	Propagation	Time Sequence	Potential Type	Waveform Character	Cellular Effects	Functions
Spike potential	Intracellular (-70 mV) Extra-cellular (3 mV)	Intracellular (synaptic) Extra-cellular (<500 μm)	Discrete recurrent	Spike – generally non-summating	Non-harmonic	Induction of subsequent spike potentials	Neural transmission
EEG rhythms	0.05–0.15 mV	Propagated through CNS	Periodic in specific sleep states	Waveform – demonstrating interference and reinforcement	Resonance – magnetic field interference	Influences neuronal tendency to develop spike potentials	No documented functions

$$\Lambda\Theta = \frac{2.3\ RT}{f}\ \log\ \frac{PK[K^+]o + PNa[Na^+]o + PCl[Cl^-]o}{PK[K^+]i + PNa[Na^+]o + PCl[Cl^-]i}$$

where

F	= Faraday's constant
P	= ion permeability constraints
K^+	= potassium
T	= degrees Kelvin
$\Lambda\Theta$	= membrane potential
Na^+	= sodium
o	= outside membrane
i	= inside membrane
Cl^-	= chloride

This complex formula demonstrates how ion concentration differences at the neuron membrane are affected by a changing external electrical potential. The formula describes the way in which extracellular oscillatory electrical rhythms such as the background frequencies of the EEG can reset neural membrane ion concentrations and affect the electrically sensitive systems of the neural cell membrane. This electrically sensitive system at the synapse and neuron cellular membrane can act as a cellular sensor to external electrical field changes, responding to changing external fields by predicating a series of changes in intracellular proteins and neuromessenger systems (Pagel, 1990, 1993a, b). Electrically induced changes in cell membrane ion concentration should affect the major cellular transducer systems. These currently poorly defined systems are potentially as complex as the extracellular neurotransmission systems at the neural synapses. Neurochemical markers of this system that we can currently measure include the G proteins, protein kinase C, and the inositol phospholipids (Gilman, 1989; Krebs, 1989). Formula 6.2 describes how changing membrane potential changes can affect cellular kinetics. This formula describes how oscillatory physiologic electrical fields such as the physiologic electrical rhythms of the EEG have the capacity to supply energy to individual cells through the production of cyclic AMP and ATP at the cellular membrane. These compounds are the primary energy repositories used chemically to provide the energy for the chemical reactions that are required for cellular functioning (Harold, 1986):

$$\Lambda Gp = n(-\Lambda\beta H^+) = F(\Lambda\Theta) - 2.3RT\ Ph$$

where

ΛGp	= kilocalorie/mole (free energy available for ATP synthesis)
$\Lambda\beta H^+$	= difference in electrochemical potential of protons inside/outside cell membrane
n	= number of protons transiting the membrane per cycle
$\Lambda\Theta$	= membrane potential
T	= degrees Kelvin
Ph	= the logarithm of the reciprocal of hydrogen ion concentration in grams atoms/liter

This formula describes the way in which intracellular energy can be transmitted across the cell membrane by frequency modulated electrical fields such as the physiological alpha, theta, and delta rhythms. This energy could be stored chemically as ATP, the primary compound supplying active energy for cellular functioning. If ATP is produced by energy derived from the external electrical field, the cell would be changed. Such energy can be considered externally derived information. This information, the manifestation of energy, could affect membrane ion concentrations, the tendency of the neuron to fire, as well as the activation of intracellular message systems. Through this system physiologic electrical fields could affect the expression of DNA – the most electrically sensitive of protein complexes and the repository of cellular memory (Pagel, 1993b, 1994a). These electrically sensitive transducer systems on the cellular membrane have the potential to utilize the EEG electrical fields to integrate the cellular memory of DNA with cognitively available sleep and waking memory. However, such a functional role for the sleep-associated electrical EEG rhythms in the process of memory remains theoretical, and the existence of functional CNS physiological electrical systems is not generally accepted. If such a functioning electrical waveform system actually exists, some of the apparently delusional contentions of psychiatric ward residents might surprisingly prove to be correct. We all may have radios in our heads that dictate some of our actions and behaviors (Fig. 6.3).

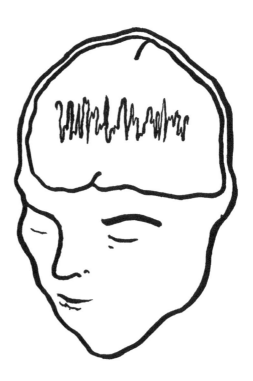

FIGURE 6.3 One view of the electrophysiological functioning of the brain.

EVIDENCE FOR A FUNCTIONING EXTRACELLULAR ELECTRICAL SYSTEM IN THE CNS

Drug Effects

Medications that produce cognitive and behavioral effects also induce changes in the EEG. Generally, psychoactive medications are found to alter the background EEG frequencies (Mandema & Danhof, 1992). In most cases, the spectrum of EEG change produced by a drug is associated with a consistent pattern of behavioral change (Table 6.2) (Herrmann & Schaerer, 1986).

CNS active drugs alter CNS electrophysiology in consistent ways. Drug class alterations of the physiologic EEG rhythms can be used to classify medications into categories and modalities likely useful in particular disease states. The way that a new drug affects the EEG can be used as a marker to predict behavioral activity, drug interactions, and potential toxicities (Itil, 1981; Pagel, 2003). Besides having consistent electrophysiological effects, most psychoactive drugs are known to exert their actions by affecting specific neurotransmitters.

Disease States

Some diseases are associated with alterations in background CNS rhythms. The technique of quantitative EEG analysis (QEEG) measures the absolute or proportional quantity of each of the physiological rhythms. A QEEG is usually obtained from a patient during the drowsy wake state with eyes both open and closed. QEEG has been proven useful as a diagnostic test for establishing or supporting diagnosis in pediatric patients suspected to have attention-deficit hyperactivity disorder (AD/HD). Pediatric AD/HD patients are more likely to demonstrate elevated levels of slow wave activity when compared to normal children (Barry et al., 2003). This elevated delta activity occurs in association with reduced amounts of higher frequency alpha and beta activity (Monstra et al., 1999). QEEG analysis of this ratio between slow and fast background EEG frequencies is generally reported as the theta–beta frequency (t-b) ratio. This ratio can be reliably

TABLE 6.2 Consistent Quantitative Alteration in Physiologic EEG Frequencies Induced by Several Classes of Psychoactive Medications

	Delta: 0.5–1.5 Hz	Theta: 5.5–8.5 Hz	Alpha: 8.5–11 Hz	Sigma: 12–16 Hz	Beta: 21–32 Hz
Benzodiazepines			↓	↑	
Tricyclic antidepressants	↓	↓			↑
SSRI antidepressants	↓		↑		
Amphetamines	↓	↓			↑
Opiates	↑		↓		
Classic Neuroleptics		↑	↓	↓	

Source: Pagel (1996).
Note ↑, ↓ – Direction of significant drug-induced change of EEG power.

utilized to differentiate pediatric AD/HD patients from normal control subjects. This method can also be used to differentiate between AD/HD subgroups, differentiating patients with symptoms that are predominately deficits in attention from those whose symptoms are predominately hyperactivity (Clark et al., 2001). Interestingly, Tibetan meditation practice has also been shown to result in QEEG changes. Highly trained monks have EEG frequency patters that are the exact opposite to the findings in these children with AD/HD. The monks demonstrate high levels of higher frequency frontal beta (gamma) frequency and much lower levels of lower frequency theta during meditation (Lutz, 2006).

Pediatric sleep apnea occurs most commonly in children with large tonsils. These children often snore, are sleepy during the day and have poor school performance. Such pediatric obstructive sleep apnea (OSA) patients often have AD/HD symptoms, with children with OSA scoring high on psychological tests for AD/HD (Chervin, 2002). Pediatric OSA produces symptoms characteristic of AD/HD that may be clinically indistinguishable from AD/HD without OSA. Sleep restriction in children can be associated with AD/HD-like behavior and problems with poor cognitive achievement. Sleepy children are far more likely than adults to demonstrate hyperactivity and attention deficit like behavior as symptoms of daytime sleepiness (Pagel et al., 2004). Pediatric OSA also induces identifiable changes in the generation of the physiologic EEG rhythms. Pediatric AD/HD patients with sleep apnea demonstrate lower amounts of alpha and have significantly reduced alpha power on QEEG analysis (Pagel et al., 2006a). Among the AD/HD patients that have abnormal theta/beta ratios (80% of the patients in some studies), the individuals with OSA have a significant reduction in EEG power for all of the physiologic frequency ranges analyzed (delta, theta, alpha, and beta) (Pagel et al., 2007). Pediatric sleep apnea consistently induces electrophysiological CNS effects that can be seen on QEEG analysis. These QEEG findings can be used to differentiate AD/HD children with OSA from those that have no apnea. The physical and psychological symptoms of pediatric OSA include AD/HD symptoms, daytime sleepiness, delays in development, social isolation, and poor school performance. These symptoms could result, at least in part, from these OSA associated changes in EEG-frequency-based CNS electrical systems (Pagel et al., 2006b).

The children with AD/HD- and OSA-induced EEG abnormalities have abnormalities of the physiological brain rhythm generation system. Since all physiologic frequencies are affected, it appears likely that these illnesses affect generalized EEG rhythm generation in the CNS. These disease states are associated with consistent abnormalities in this system indicating that the EEG frequencies are likely to reflect a functioning system in the CNS. Yet, most neuroscientists have yet to acknowledge the existence of this system. Clearly, further advances in our knowledge of neural electrophysiology are one key likely to lead to an expanded knowledge of cognitive states.

The human CNS is likely the most complex of physiologic systems that we have tried to understand. Neurochemistry and neuroanatomy, though complex, are components of this system that current science can clearly describe. Yet these components may be outweighed in both importance and function by other systems. What little we understand of CNS electrophysiology suggests that a fascinating and entire system of neurocognition functions outside the awareness of the neuroscientific mainstream.

THE ELECTROPHYSIOLOGY OF DREAMS

Dreams occur during the stages of sleep. We define the sleep stages based on the presence or absence of background electrophysiologic rhythms in the recorded EEG. Dreams are commonly reported from sleep stages that are characterized and defined by their associated electrophysiologic rhythm. These electrophysiologic rhythms can potentially function as neurosignals supplying energy to cells, affecting membrane ionic equilibrium and intracellular energy stores. These physiologic electrical rhythms have the potential to alter and affect assess to the cellular memory inscribed in DNA. Medication-induced alterations in this system and disease states known to affect this system are known to alter our cognitive interaction with our environment. It may be that dreams are our conscious feedback, our personal CNS message systems, giving assess to the current status and level of functioning of this system. This is an unproven, yet apparently reasonable hypothesis. But it is another possible theory in the history of grand theories of dreaming.

Each stage of sleep is associated with characteristic background EEG frequencies. There is clear and consistent evidence that dream recall frequency is different from the various stages of sleep (Foulkes, 1985; Goodenough et al., 1965). This difference in recall may, however, reflect the distance of each sleep stage from waking rather than the underlying sleep stage and the EEG frequency defining that stage (Koukkou & Lehmann, 1983). There is some evidence that dream recall and intensity increases during periods of alpha activity in REMS (Ogilvie et al., 1982). Dream content may also differ based on the sleep stage from which the dreamer awakens. REMS dream content is postulated to be more bizarre and hallucinatory than the dreams reported from other sleep stages (Kahn & Hobson, 2005). However, the small amount of experimental work on which this postulated association is based is limited and troubled by methodological problems. The difference in dream content that has been found between sleep stages is associated with the length of the dream report. When the length of the report is controlled, differences in dream content between sleep stages tend to disappear (Domhoff, 2003).

There are a series of diagnoses that induce changes in these background EEG rhythms. As discussed earlier in this chapter, both AD/HD and OSA change the power and pattern of EEG frequency generation in children. In adults, there are a group of diagnoses known to increase alpha frequency intrusion into sleep including primary insomnia, fibromyalgia, chronic pain, and chronic fatigue syndrome. Patients with such alpha intrusion present on EEG often complain of non-restorative sleep despite sleep that is otherwise normal in both duration and quality on polysomnography. None of these diagnoses are known to be associated with changes in dream recall or content (Pagel & Shocknassee, unpublished data). Alpha intrusion also occurs in individuals taking the selective serotonin reuptake inhibitor (SSRI) antidepressants such as Prozac, Zoloft, Paxil, Lexapro, and Celexa. Dream content and frequency are known to change during recovery from depression with dreams occurring more frequently and with increased "positive" content, but it is unclear whether the EEG changes induced by these medications have anything to do with dreaming (Pagel, 2004; Pagel & Pandi-Perumal, 2007).

There has been little work done in this new field of cognitive electrophysiology and further research is likely to totally change our admittedly limited understanding of brain electrophysiology. The primary electrophysiologic correlates of dreaming remain the various sleep stages. Dream recall occurs at highest frequency and intensity on arousal from REMS and sleep onset. Although sleep onset is strongly associated with alpha frequency, and REMS has a strong association with theta, these sleep stages are also those which are closest to waking. This association, loose and poorly defined, is currently the best that electrophysiological neuroscience has to offer the field of dreams.

CHAPTER 7

The Complexity of Dreams: Neural Networks and Consciousness

I don't know whether you have ever seen a map of a person's mind. Doctors sometimes draw maps of other parts of you, and your own map can become intensely interesting, but catch them trying to draw a map of a child's mind, which is not only confused, but keeps going round all the time. There are zigzag lines on it, just like your temperature on a card, and these are probably roads in the island; for the Neverland is always more or less an island, with astonishing patches of color here and there, and coral reefs and rakish-looking craft in their offing, and savages and lonely lairs, and gnomes who are mostly tailors, and caves through which a river runs, and princes with six elder brothers, and a hut fast going to decay, and one very small old lady with a hooked nose. It would be an easy map if that were all; but there is also first day at school, religion, fathers, the round pond, needlework, murders, hangings, verbs that take the dative, chocolate pudding day, getting into braces, say ninety-nine, threepence for pulling out your tooth yourself, and so on; and either these are part of the island or they are another map showing through, and it is all rather confusing, especially as nothing will stand still.

(Barrie, Peter Pan & Wendy, 1911/1988)

The advent of the "thinking machine," the computer, has led scientists theorize how the human brain might function as a biologically based computing system. Neurons can be conceptually viewed as logical on–off switches with the capacity to function like a digital binary computer system (McCullough & Pitts, 1943). These mathematicians postulated that the central nervous system (CNS) functions as a massively parallel system of interconnected processing elements called neurons. The human CNS is comprised of at least a hundred billion neurons, each with multiple synaptic connections. The level of complexity of such a system became quickly obvious to scientists developing neural network models of such a system (Figure 7.1).

In order for such an incredibly large and complex neural net to function, an organizational system for neural interconnections is required. In 1948, Donald Hebb proposed in his book *Organization of Behavior* that the firing of synaptic inputs leads to an increase in the strength of the neural connections between neurons. Hebb wrote,

> The most obvious and I believe much the most probable suggestion concerning the way that one cell could become more capable of firing another is that synaptic knobs develop and increase the area of contact between [nerve cells] (1949, p. 62).

If a system such as this exists, repeated exposure to external stimuli could alter the pattern of activity of a neural network or "net." Based on this schema, the

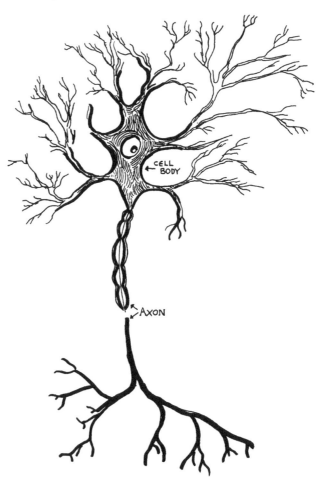

FIGURE 7.1 An example neuron.

potential exists for a complex random system such as the neural net of the CNS to settle down into a specific activity pattern, progressively evolving toward a more stable system with the ability to respond quickly in a similar manner to repeated stimuli (Hinton et al., 1984).

Artificial interconnected systems have been developed using conceptual "on–off" neural net neurons. The functional, highly distributed, parallel processing systems developed on this concept tend to be fault tolerant. Some neural net systems have been found to have applications in digital computation systems, addressing computational problems that are poorly amenable to linear mathematical approaches. Neural nets have also been proposed as the basis for much of the functioning of the CNS. Neural nets have even spawned a theory of dreams – the dream as a ghost in the machine.

BIOLOGICAL APPLICATIONS OF NEURAL NET THEORIES

Neural net theories of parallel processing offer conceptual advantages over classic transmission line, hardwired neural connection theories of nerve cell function. If one individual neuron quits functioning, the neural net system continues to perform without significant error. As more elements "die," there is a gradual drop off in performance, much as would be seen in a biological system (Coveny & Highfield, 1995). The plasticity of interconnected neural networks would allow the brain to self-organize networks for recognition of perceptual input. Repeated perceptual input could be organized into a set of neural interconnections that allow the structure of the brain to make connections to external world objects and events. In rabbits micro-electrode studies in the hippocampus (an area closely associated with memory) indicate that repeated artificial neural stimulation can increase synaptic strength. Evidence suggests such changes that can last for months. These nerve nets may be brain-based correlates for the incorporation of experience into memory (Bliss & Lomo, 1973). Larger computer models involving over 10,000 simulated neurons have been developed. Such systems are prone to the development of population oscillations in electrical activity. Interestingly, this oscillatory electrical activity resembles the theta wave patterns that occur in association with rapid eye movement (REM) sleep (Traub & Miles, 1992).

THE DREAMS OF NEURAL NETWORKS

Crick and Mitchinson (1983, 1995) propose that dreams may be a natural result of neural net overloading in the CNS. During sleep when the neural net is disconnected from normal inputs and outputs, they propose, CNS nets produce outputs that are combinations of stored associations. The system utilizes random input to repetitively weaken the associative connections connected with overload associated with perceptual input during waking. In this way the CNS undergoes a process of reverse learning that eliminates extraneous associative memories and neural connections. This theory purports to account for the association of REMS dreams with random and bizarre intrusions of previously stored memories. Based on this theory, the dream is a meaningless mishmash of associations arising from an overloaded network, neural connections that need to be unlearned. Dreams can be considered the screensavers of our personal computing systems. Dreams – the non-conscious, unfocused ramblings of our consciousness computer trying to clear itself of excess data and poorly developed programs in preparation for the next day's functioning. From this perspective, dreams have little meaning, except as reflections of the extent of disorder present in our waking and sleeping consciousness.

COMPLEXITY OF THE BIOLOGIC SYSTEM

The hundred thousand million neurons in the CNS of one human is a number similar to that for the number of stars in our galaxy. These neurons have thousands of millions of synaptic connections, each with the capacity to respond to multiple neurotransmitters. And that is just the connected neurotransmission system. There are neural cells responding to generally secreted neuroendocrine chemicals as well. These chemicals, such as melatonin and thyroid stimulating hormone, can alter both specific and general areas of the CNS affecting neural response to stimulus and neuron metabolism. There are electrical gap junctions utilized by many of the other millions upon millions of supporting brain glial cells to communicate with neighboring neurons through direct electrical connections at the cell membrane, a neural communication system that operates outside the chemical neurotransmission system. There are also the background extracellular electrical fields that have the capacity to affect the expression of the complex cellular memory and information system stored in DNA (Chapter 6). The different systems functioning in waking cognition utilize multiple and complex junctions integrating anatomic structures with their physiological and psychological functions. The human CNS may very well be the most complex system that we have attempted to understand. That complexity is evidently required in order to represent and map an internal representation of external reality, making an orderly presentation of the seemingly chaotic experience of living in this world.

Some theorists go further in arguing that increasing complexity may be a goal of living systems. Paul Davies postulates that life and consciousness are emergent phenomena of complexity (2006). At a certain level of complexity, conscious self-awareness occurs. Davies postulates that the minimal complexity required for consciousness on a quantum level may be at the level of 400 entangled particles. Even such a small system has an inherent complexity of possible alternative states that rivals the number of particles existing in the known universe (10 to the 80th power). The science of complexity has brought us once more to the border between body and mind. The implication is that at a certain level of complexity, a system is no longer just mechanical, biological, or cosmological. In order to interact with the external environment, the system must self-organize, take on the attributes of mind, and become conscious.

THE LIMITS OF DREAM

Structure

The hard neurosciences have been utilized to study dreaming, with the neuroanatomy of dreaming suggesting that mind is a global process not restricted to the brain stem or the pineal gland as Descartes once suggested. Major CNS trauma to the prefrontal cortex, what was once prescribed as psychosurgery for untreatable psychiatric patients, results in a loss in dream recall. Depression and obstructive sleep apnea, and illnesses with global effects on the CNS are the disorders that affect dream recall.

The neurochemistry of dreaming indicates that the same drugs that affect our waking consciousness affect our dreaming. It appears that dreaming is not a simple state triggered by acetylcholine, modulated by limited other neurotransmitters that are affected by only a few medications. Dreaming is a state affected by the same spectrum of neurochemicals that shape waking consciousness.

Studies of electrophysiology indicate that dreams occur in association with the global electroencephalogram (EEG) frequency patterns that characterize sleep. These physiological electrical rhythms have the capacity to affect neural function outside the sequential spike potential cascade utilized in synaptic neuron-to-neuron communication. These electrical potentials can set cellular equilibrium, supply energy, and affect the expression of DNA, acting as neurosignals in the CNS.

The Neurological Substrate of Dream

We have taken each of these neurostructural sciences to their limits in attempting to understand how each science attempts to explain the neurological substrate of dreaming. Some work has been misleading. Other work has been wrongly used to provide evidentiary support for grand but untenable theories of dream. But scientifically based research into its basic neurological nature suggests that dreaming is likely to be based on the following structural characteristics:

1. Dreaming is not controlled by a localized or specific structure somewhere in the brain. REMS is one of the dream-associated states that is likely to have a neurological on–off switch utilizing the neurochemical acetylcholine at specific brain stem sites. But the mental activity occurring during sleep that we call dreaming does not have a detectable on–off neurological switch. *Dreaming is a global process.*

2. Dreaming is a state of consciousness arising diffusely out of the global state of sleep in the CNS. *Both sleep and dream appear to utilize the same structural framework of neurochemicals and cellular neuroanatomy as our waking consciousness.*

3. Research into the limits of neural electrophysiology suggests that processes other than classical synaptic neurotransmission are likely active components of CNS function. *These processes have the capacity to supply energy, set equilibrium, to signal individual neural cells, and to affect the expression of DNA.*

4. Classical and simplified models of dreaming that equate REM sleep with dreaming are outmoded and based on belief rather than evidence. *REMS and dreaming are doubly dissociable.*

5. In the scientific study of dreaming, complex rather than the simplified models better describe the data. *Dynamic models of complex parallel neural processing systems provide models likely to be useful in our attempts to understand the mind /brain interface.*

6. The cognitive processes that are often referred to as mind, such as associative thought, reasoning, creativity, belief, and consciousness, are poorly explained

by neuroanatomy, neurochemistry, and electrophysiology. *Mind-based cognitive processes including dreaming are among the most complex systems we have yet attempted to understand.*

7. It may be that the level of complexity of perception and nervous system processing requires that something like mind develop in the organism, something with the capacity to integrate the unimaginably complex perceptual input we receive and to interpret the output of our interactive processing modalities into a functional coherence. *The cognitive processes that are mind based can be defined as those processes that are at a level of unexplainable complexity.*

Simple answers and simple models no longer seem to work for the hard sciences addressing CNS structure. The concept of the CNS as a series of connected transmission lines between brain sites works to provide a limited model of concrete brain and neuron functioning. But at the interactive border of cognitive neuroscience, neuroanatomy, neurochemistry, and electrophysiology mingle in an area of increasing complexity. In cosmology, mathematics, and quantum mechanics the application of concepts of complexity has often served to mark the limits of our capacity for intellectual understanding of that area of study. The incredible complexity of the CNS neural net systems may herald our arrival at that same limit marking the border between brain and mind. It is a fuzzy border. Here is a place where we can see how much there is that we still do not understand about brain function. At the extremes of the basic neurosciences we reach a place just outside the eclipse horizon of mind. Limited by our current mental, perceptual, and technical capacities, it is a border difficult for us to see across even when we try to use the window provided by dream.

Studying the Cognitive Substrate of Dreaming

Dreams are more than their structural basis. Dreaming includes a group of diverse, interrelated cognitive functions. A dream is a cognitive process, a process by which information is manipulated in the central nervous system (CNS). The process can be affected and influenced at different levels. These levels include whether a dream is remembered (recall), the storyline of the dream (content), and dream incorporation into waking behavior (use). Different factors can affect the dream at each processing level.

We can study the dream as a memory – a memory of thoughts that occurred during sleep. The study of memory is the cornerstone of cognitive science, and we understand memory much better than we understand dream. Dream as a process differs from most memories in having only minimal perceptual (sensory) input and none of the encoding process required for that perceptual input. Types of non-perceptually based memory similar to the memory of dreaming include imagery and intrinsic memory. Imagery is a type of memory process in which new visual representations are created without utilizing perceptual input. Intrinsic memory, the memory of basic concepts, is also not perceptually based. Both imagery and intrinsic memory can be studied at the levels of transfer into conscious awareness, and at the level of organization and expression. Dreaming is likely to utilize the same neural processing systems as non-perceptually based memories in its integration into conscious thought.

In order for a dream to influence conscious thought, the dream must first be remembered or recalled while we are awake. Once we recall a dream, we develop associations both intrinsic and extrinsic to that image before trying to understand the meaning of the dream. At this point, many of us will share the dream either verbally with another individual or by writing the dream down in a journal. The dream becomes available for outside interaction and interpretation. It is at this point that a dream can be studied (Cavallero et al., 1990).

We can study dream recall frequency using questionnaires, nocturnal awakenings of dreamers, and by patient diaries of dream reports. There is a remarkably high correlation between the results obtained utilizing different assessment methods for dream recall frequency (Schredl, 2002). On average, we recall a dream somewhere between once and twice a week (Pagel & Vann, 1992).

We can study reports of dream content. This has been the focus of dream studies for millennia. The study of dream content involves at least two memory processes: the associations of this information with other stored memories (ecphory) and the organization of this material for expression. Content studies require that the dream be written or described to a third party. If the dream is described to another person, the interaction between the dreamer and the individual recording the dream can affect the described report of dream content. Both repression and transference can affect this relationship and affect reports of dream content (Ekstrand, 1977). Studies have demonstrated that dream content is altered by personality, gender, social background and ethnicity, concomitant psychiatric illness, stress and trauma, and by daily life events (Moffitt et al., 1993).

The study of memory also suggests that dreaming can be studied at the level of incorporation into behavior. This approach has only recently been applied to dreaming (Kuiken & Sikora, 1993; Pagel & Vann, 1992). Dream use, the incorporation of dreaming into behavior, is altered by gender, age, dream recall, and obstructive sleep apnea (Pagel & Vann, 1992, 1995). Curiously, stress and ethnicity have been shown to have minimal effects on dream incorporation into behavior (Pagel & Vann, 1993, 1995).

The cognitive process of dreaming requires the cognitive substrates of sleep and memory. During sleep the complex interplay of neural networks has been postulated to approximate a form of consciousness during dreaming. Artificial intelligence (AI) systems can be designed in an attempt to approximate brain-based cognitive functioning. The current capabilities of AI systems utilize artificially created cognitive processes including perception, calculation, communication, and motor skills. In these areas, AI has demonstrated capabilities far beyond those of biological systems. However, at this point in time there is little indication that AI systems can be created that have aspects of mind. But the creators persist, fashioning more complicated systems using massively parallel neural nets that approximate brain functioning. There is always the possibility that AI can be taught to dream.

CHAPTER 8

Dreams and Sleep: The Substrate Hypothesis

To sleep: perchance to dream ...

(Shakespeare W. Hamlet III, I, 36)

A substrate is the foundation or base that a biologic system requires for its existence. In order for dreaming to occur, an underlying neurophysiological substrate must be present. The cognitive substrate required for dreaming, sleep, is required in order for sleep-associated mentation (dreaming) to occur. The minimum requirements for dreaming include sufficient sleep in order for dreaming to take place, and at least some uninterrupted periods of the sleep stages primarily associated with dreaming (Pagel & Shocknesse, 2006). As the quality of the sleep substrate declines, as sleep becomes increasingly fragmented and disturbed, the number of dreams remembered on waking declines as well. Polysomnography (PSG) is the technical procedure currently used to quantitatively assess sleep. PSG variables in sleep quantity or quality that significantly affect reported dream recall frequency are altering the substrate of sleep required for dreaming to occur (Table 8.1).

There has been little investigation into this area and researchers have tended to ignore the process of sleep and focus on the association of rapid eye movement (REM) sleep with dreaming. Recently, some studies have begun to address the affects of sleep-associated diagnoses on the recall of dreaming. Dream recall frequency is lower in patients with depression (Armitage et al., 1995), obstructive sleep apnea (OSA) (Pagel & Vann, 1995), increased daytime sleepiness (Myers & Pagel, 2001), frontal–parietal central nervous system (CNS) damage (Solms, 1997), and periodic limb movement disorder (PLMD) (Schredl, 2001). These pathologies disrupt sleep and reduce the quality of the experience of sleep in ways that are specific for each disease process. These diagnoses share the characteristics of inducing poorer sleep quality as well as a reduced frequency of dream recall.

The diagnoses most commonly associated with disordered sleep are those that cause insomnia. Insomniacs have difficulty with sleep initiation, sleep maintenance,

67

FIGURE 8.1 A sleeper.

or have the complaint of non-restorative sleep. For insomniacs, disordered sleep is associated with daytime impairment. Definitions for insomnia range from "dissatisfaction with sleep" to structured definitions that require the objective PSG defined presence of nocturnal sleep disturbance. Insomnia is generally considered a complaint or "symptom" rather than a diagnosis (Pagel, 1994). When symptoms are present for at least 3 months an individual can be considered likely to have one of the several diagnoses causing chronic insomnia (Edinger et al., 2004). As seems typical in the area of cognitive science, the use of varying definitions has led to confusion as to the general prevalence, risk factors and morbidity for "insomnia" (Ohayon, 2002). However, most studies suggest that approximately 15% of the population has a chronic every night problem with insomnia (Hohagen et al., 1993).

When we study insomniacs in the sleep laboratory, we often find typical PSG markers for insomnia, including increased sleep latency (SL) which is the time between attempted sleep initiation and sleep onset, increased awakenings and arousals, decreased sleep efficiency (SE), increased time waking after sleep onset (WASO), and decreased total sleep time (TST). Many insomniacs will also have specific sleep stage reductions in REM sleep and deep sleep (Stage 3–4) (Chervin, 2003; Leister, 2005). Although these findings are present in some insomniacs, not all individuals complaining of insomnia have objective abnormalities in sleep when studied in the sleep laboratory. This lack of PSG findings indicates that insomnia is often a patient complaint rather than an objective finding. There is no true diagnostic test for insomnia.

DREAM RECALL IN INSOMNIA

At least half of sleep laboratory patients complain of insomnia. In a study done at our sleep laboratory in Colorado, we found that the complaint of insomnia is not associated with a significant difference in dream recall. Nightmare frequency is significantly higher in patients complaining of insomnia (Pagel & Shocknesse, 2006) (Table 8.1). When we looked at the sleep variables obtained from the PSGs

TABLE 8.1 Comparison of Means of Sleep Laboratory Variables Between Group Reporting Insomnia and Group Reporting No Insomnia (N = 289)

	#	Age	Gender (F = 1, M = 2)	Dream Recall (1–5)	Night-mare Recall (1–5)	TST (Min-utes)	SL (Min-utes)	REMS Late-ncy (Min-utes)	WASO (Min-utes)	Arousal-Wake Index (#/hour)	REMS (Min-utes)	Stage 1–2 (Min-utes)	Stage 3–4 (Min-utes)	AHI	PLMI	Ep-worth
No Insomnia																
Total no Insom-nia group	*148*	*49.2*	*1.54*	*2.68*	*2.07**	*294.1*	*23.3*	*143.4*	*68.5*	*15.9*	*45.9*	*225.2*	*23.0*	*15.3*	*15.0*	*10.3*
Not taking psychotropic medications	97			2.70	2.06*	293.5	22.2	125.3	70.5	13.9	48.1	209.5	25.9	14.3	16.0	9.4
Psychotropic medications – current use	51			2.73	2.09*	295.0	25.6	172.5	65.0	19.9	42.2	234.8	18.0	17.0	13.3	11.8
Primary no Insomnia group	55	45.5	1.52	2.90*	2.26	318.7	18.9	123.0	54.6	5.04	51.8	233.9	33.0	7.75	5.77	10.4
Insomnia																
Total Insomnia group	*141*	*45.9*	*1.46*	*2.66*	*2.38**	*302.7*	*24.1*	*145.5*	*60.9*	*15.3*	*43.8*	*239.2*	*19.7*	*12.3*	*14.8*	*10.3*
Not taking psychotropic medications	72			2.60	2.31*	318.8	19.5	135.2	64.0	15.2	49.6	250.0	19.2	13.2	15.2	10.6
Psychotropic medications – current use	69			2.73	2.46*	291.2	28.7	157.4	57.0	15.4	37.7	233.2	20.3	11.5	14.5	9.86
Primary Insomnia group	49	45.2	1.55	2.52*	2.35	321.5	19.9	132.4	58.0	9.79	51.3	249.6	20.6	7.75	5.35	10.1

AHI: apnea–hypopnea index.
Significant difference between similarly classified groups
*denotes statistical significance >.05

done on these patients, we found a series of variables that were often abnormal in the patients that had lower levels of dream recall. The sleep latency (*SL*) is the time required for an individual to fall asleep after going to bed in order to attempt sleep. This value is prolonged (>15 minutes) for individuals with sleep onset insomnia. We found that for individuals with a prolonged SL (>30 minutes) there was a significant decline in dream recall. Sleep efficiency (*SE*) is a measure of the amount of time in bed that one is actually asleep. In this study, individuals with normal SE ($>85\%$), and those with a moderate decline in efficiency ($65–85\%$) had

TABLE 8.2 The Effects of PSG Sleep Variables on Reported Dream and Nightmare Recall Frequency, Total Study Group ($N = 289$)

	Dream Recall (1–5)	Nightmare Recall (1–5)
SL (minutes)		
<5.0 (#48)	2.7 + 1.02	2.3 + 1.04
5–15 (#107)	2.7 + 0.93 (b)	2.3 + 0.96
15–30 (#76)	2.8 + 0.87 (c)	2.3 + 0.99
>30 (#58)	2.4 + 0.92 (d)	2.0 + 0.97
	$p < 0.05$ d < b;	
	$p < 0.01$ d < c	
TST (minutes)		
<300 (#127)	2.6 + 0.96	2.1 + 0.98 (d)
300–360 (#103)	2.7 + 0.96	2.3 + 0.98
>360 (#59)	2.8 + 0.84	2.4 + 0.99 (f)
		$p < 0.05$ f > d
SE		
<65% (#69)	2.4 + 0.97 (a)	1.9 + 0.78 (d)
65–85% (#115)	2.7 + 0.75 (b)	2.1 + 0.88 (e)
	$p < 0.05$ b > a	
>85% (#105)	2.8 + 0.86 (c)	2.5 + 1.07 (f)
	$p < 0.01$ c > a	$p < 0.01$ f > d;
		$p < 0.02$ f > e
WASO (minutes)		
<30 (#81)	2.9 + 0.90 (a)	2.6 + 1.12 (d)
30–60 (#76)	2.8 + 0.94 (b)	2.3 + 0.86
>60 (#132)	2.5 + 0.91 (c)	2.0 + 0.90 (f)
	$p < 0.05$ a > c, b > c	$p < 0.02$ d > f
Arousal-wake index (#/hour)		
<5 (#109)	2.7 + 0.92	2.3 + 0.99
5–15 (#90)	2.8 + 0.94	2.3 + 0.92
>15 (#90)	2.5 + 0.93	2.0 + 1.00
REMS (minutes)		
0 (#23)	2.5 + 0.93	2.2 + 0.84
1–30 (#67)	2.6 + 1.22	2.1 + 1.07 (f)
30–60 (#114)	2.6 + 0.73	2.2 + 0.97
>60 (#85)	2.8 + 0.73	2.4 + 0.87 (h)
		$p < 0.05$ f < h
Stages 3 and 4 (minutes)		
0 (#125)	2.7 + 0.82	2.2 + 0.87
0–30 (#85)	2.6 + 0.83	2.2 + 0.83
>30 (#79)	2.7 + 0.86	2.3 + 1.08

significantly higher dream recall frequency than the grouping with very poor sleep (SE < 65%). Wake after sleep onset (*WASO*) is a measurement in minutes of the time awake after sleep onset. Normal individual spend less than 30–60 minutes of time awake after initiating sleep. The individuals in this study with a *WASO* of more than 60 minutes reported significantly lower dream recall than individuals with a *WASO* in the normal range (Table 8.2).

In a clinical sleep laboratory more than 80% of patients will have obstructive sleep apnea (OSA). In individuals with OSA (defined as more than 15 episodes of breathing disturbance per hour) insomnia affects dream recall in a manner opposite to what is found in the other groups. In the total group, the presence or absence of the complaint of insomnia makes minimal difference in dream recall frequency. For individuals with OSA, however, dream recall is significantly higher for the insomniacs. Increasing daytime sleepiness as reported using an Epworth Scale (a self-reported rating of daytime sleepiness) leads to a decrease in dream recall. Periodic limb movements (as rated by a PLM index (PLMI) of number of events per hour) has no effect on dream or nightmare recall in this study (Table 8.3).

This study indicates that individuals with lower PSG sleep quality have lower dream and nightmare recall. SL had the greatest effects on dream recall frequency. This finding could indicate that these patients who have difficulty getting to sleep have less sleep onset dreaming. Like REMS, sleep onset is a sleep stage with high levels of reported dream recall on awakening. But a prolonged SL also indicates difficulty in transitioning from wake to sleep. It is possible that dysfunctions in this transition may negatively affect the memory processes involved in the waking recall of dreaming.

Dream recall is also lower for individuals who have difficulty with maintaining sleep after sleep onset. Individuals with poor SE, at a level less than 65%, reported lower levels of dream recall. When time WASO was greater than 60 minutes there was also a decline in dream recall. These findings indicate that individuals with

TABLE 8.3 The Effects of Apnea–Hypopnea Index (AHI), PLMI, and Epworth on Reported Dream and Nightmare Recall Frequency, Total Study Group (*N* = 289)

	Dream Recall (1–5)	Nightmare Recall (1–5)
AHI (#/hour)		
<5 (#50)	2.9 + 0.86	2.4 + 0.97
5–15 (#143)	2.6 + 0.73	2.2 + 0.82
>15 (#96)	2.6 + 1.04	2.1 + 1.13
PLMI (#/hour)		
<5 (#102)	2.7 + 0.78	2.4 + 0.96
5–15 (#98)	2.8 + 0.81	2.2 + 0.98
>15 (#89)	2.5 + 0.98	2.2 + 0.91
Epworth (0–24)		
<5 (#42)	2.5 + 0.72	*1.9* + 0.63 (e)
5–10 (#73)	2.6 + 0.77	2.2 + 0.95
10–15 (#41)	2.9 + 0.81	*2.6* + 0.97 (f)
		$p < 0.01$ f > e
>15 (#46)	2.9 + 0.99	2.4 + 1.12

Note: (+) denotes standard deviation from mean value.

poor sleep through the night have less dream recall. It appears that a certain quality of sleep is required if dreaming is to occur. As the quality of the sleep substrate declines, dream recall declines as well.

INSOMNIA AND NIGHTMARE RECALL FREQUENCY

Different PSG sleep variables affect nightmare recall than affect dream recall. Nightmares are different from ordinary dreams. They are defined as vivid and distressing mental experiences generally associated with REM sleep and often causing an arousal from sleep. Typically, nightmares are coherent dream sequences that seem real and become increasingly more disturbing as they unfold. Associated emotions usually involve anxiety, fear, or terror but frequently also anger, rage, embarrassment, disgust, and other negative feelings. Dream content most often focuses on imminent physical danger to the individual such as the threat of attack, falling, injury, or death but may also involve aggression toward others, potential personal failures, suffocation, and other distressing themes.

Nightmares are diagnostically classified as a REM sleep parasomnia, with parasomnias considered to be undesirable physical events or experiences that occur during entry into sleep, within sleep, or during arousals from sleep. Parasomnias include sleep-related movements, autonomic motor system functioning, behaviors, perceptions, emotions, and dreaming. These are sleep-related behaviors and experiences over which the sleeper has no conscious deliberate control. Parasomnias become clinical diagnoses when associated with sleep disruption, nocturnal injuries, waking psychosocial effects, and adverse health effects. Individuals with the diagnosis of nightmare disorder have frequent frightening dreams that disturb sleep and affect daytime performance. This diagnosis may be associated with personality characteristics including fantasy proneness, psychological absorption, dysphoric daydreaming, and "thin" boundaries. Frequent nightmares can be associated with psychopathologies such as schizophrenia-spectrum disorders, anxiety disorders, and dissociative disorders. Complaints of recurrent nightmares are higher in individuals that score poorly on tests designed to measure an individual's perceived quality of well-being (Besiroglu et al., 2005). An association of such psychopathology has been detected for adults and adolescents but research on children is largely absent.

There is no definite agreement between studies on the prevalence of nightmare disorder. Ten to fifty percent of children aged 3–5 have nightmares severe enough to disturb their parents. A larger percentage, probably 75%, can remember at least one or a few nightmares in the course of the childhood. Approximately 50–85% of adults admit to having at least an occasional nightmare; 2–6% of college students and 4–6% of adults report one or more nightmares per week (Levin, 1994). From 5% to 30% of children are reported to "often" or "always" experience nightmares, with that high frequency persisting in female adolescents and college students (Belicki et al., 1997). There is a gender difference in nightmare frequency (female > male) that persists into adulthood (Wood & Bootzin, 1990). From 2% to 8% of the general population has a current problem with nightmares. This frequency can be much higher in psychiatry clinic populations (Krakow et al., 2001).

When there is a history of significant physical or psychological trauma, recurrent nightmares are likely a symptom of posttraumatic stress disorder (PTSD).

Nightmares are the most common symptom of PTSD, a disorder characterized by symptoms of "hyper-arousal" occurring after serious physical or mental trauma. The severity, duration, and proximity of an individual's exposure to the traumatic event are the most important factors affecting the likelihood of developing PTSD. Social supports, family history, childhood experiences, personality variables, and pre-existing mental disorders may affect the development of PTSD. However, this disorder can develop in individuals without any predisposing conditions, particularly if the stressor is extreme. PTSD is extremely common in some populations such as displaced immigrant groups where more than 50% of the individuals may have PTSD. PTSD can be diagnosed in from 3% to 58% of combat veterans and victims of major trauma (Pagel & Nielsen, 2005).

Nightmares and negative dream reports are more common in patients with insomnia (Ohayon & Morselli, 1994). This may be because dream salience or intensity increases the potential for recall of a particular dream (Kuiken & Sikora, 1993). Such salience and intensity is the characteristic of the nightmare. A decline in nightmare recall occurs as sleep becomes increasingly disordered. Individuals with a prolonged SL (>30 minutes) report a significant decline in nightmare recall. Nightmare recall is more common for individuals with more than 6 hours of total sleep time (TST) for the night of study when compared to those with a TST of less than 5 hours. Individuals with better than 85% SE have higher nightmare recall frequency than individuals with poorer SE, and individuals with less time awake after sleep onset (WASO < 30 minutes) have more frequent reports of nightmares (Pagel & Shocknasse, 2006). The PSG data demonstrates that reported nightmare recall frequency, like dream recall frequency, declines as PSG sleep quality declines. These findings indicate that like dreaming, nightmares require a certain level of sleep quality in order to occur (Tables 1–3).

However, different sleep variables affect nightmares than affect dream recall. Nightmare recall is less in individuals with lower amounts of REMS. A decrease in REM sleep time (<30 minutes) is associated with a significant decline in nightmare recall frequency. This finding, not present for dream recall, might be expected since dreams are often reported from other stages of sleep than REMS. Nightmares, with their negative emotions, content focusing on imminent physical danger, and other distressing themes and arousal from and disruption of sleep, are a REM sleep parasomnia. The nightmare is a typical type of REM sleep dream.

THE ASSOCIATION BETWEEN SLEEP AND MIND

We spend one-third of our lives sleeping, yet it has been difficult to demonstrate that sleep has clear functions. We know that even severe and prolonged sleep deprivation does not result in death or acute illness. Such horrible studies were done in the Nazi death camps. The negative impact of sleep deprivation is on waking function. An individual deprived of sleep, even for days at a time, typically becomes unable to stay awake and has short repetitive micro-sleeps while awake. Performance declines on cognitive tasks that are boring or repetitive. After prolonged sleep deprivation, some individuals will experience visual or auditory hallucinations. These are the types of performance dysfunctions associated with sleep deprivation that have been in part responsible for such disasters as the Exxon Valdez wreck, the Three-Mile Island disaster, and the crash of the Challenger.

Sleep deprived individuals are much more likely to be involved in fatal autocrashes (Knipling & Wang, 1995). Severe insomniacs with chronic sleep deprivation are also prone to the development of mood disorders such as depression and when stressed more likely to attempt suicide (Bernert et al., 2005). Perhaps surprisingly, such sleep deprived individuals maintain their ability to respond to emergencies. Some individuals have shown their capacity to function in what appears to be a normal fashion when averaging less than 4 hours sleep per night. But this is unusual, and most individuals chronically deprived of sleep report a decline in quality of life, and have a decline in waking function that can be detected on cognitive tests (Breslau et al., 1996).

It was Freud's perspective that dreamless sleep was physiologically superior to dreaming sleep: "There ought to be no mental activity in sleep; if it begins to stir, we have not succeeded in establishing the foetal state of rest" (1917). Freud's postulate was that dreams functioned as guardians of sleep, getting rid of potential disturbances that could interfere with the organism's attempt to attain restorative levels of dreamless sleep. In the physiology laboratory, it has been difficult to show restorative effects for either dreamless or dreaming sleep on neurochemical or other metabolic systems. REMS deprivation can affect incorporation of learned information into memory. However, many of the patients studied in the sleep laboratory have little if any REMS and appear to function well in society. Current theories of sleep function have reached back to evolutionary theory, postulating that sleep during darkness had survival value for the developing species beset by prowling carnivores on the plains of Africa (Revonsuo, 2003).

It is interesting to postulate that one of the functions of sleep may be to allow dreaming to occur. That is of course if dreaming has a function. The function of dreaming is a matter, as noted elsewhere in this book, of some debate. It is less debatable that dreams do provide some sort of assess, a window perhaps, into the functional levels of our CNS that are poorly described by current techniques used to assess CNS structure and function – what we have suggested before may be the mind. In that case, the requisite function for dreams, and the sleep required in order for us to have dreams, may be to access mind.

CHAPTER 9

Dreaming and Memory

Memory is the diary that we all carry about with us.
(Oscar Wilde, *The Importance of Being Ernest*, 1895)

A dream is a cognitive process, and, like other cognitive processes, it can be affected and influenced at different levels of cognition. One way of considering the dream is as a memory – a memory of thoughts that occurred during sleep. Using this approach, we can apply to the study of dreaming the insights that we have obtained into the cognitive modeling of memory.

The processing of memory is a requirement for human functioning. Individuals experiencing the loss of the ability to recall memories are often unable to function normally. Because of this, dysfunctions of memory are considered illness. Because of these attributes, memory has been more rigorously studied than dreaming. Neuroanatomical correlates of memory are centered primarily in an area of the brain called the hippocampus and can be disrupted by focal central nervous system (CNS) damage to that area. Electrical stimulation of specific sites of the hippocampus during neurosurgery in awake patients can result in the reporting of detailed memories not previously recalled by the subject of the surgery. There is evidence that learning and the recall of both short- and long-term memories are components of memory based on specific anatomic CNS sites. The neurological sites associated with the immediate recall of memories and memory retrieval are less well defined.

Memory is not just the model of a cognitive process. We each have the everyday experience of memories of our personal experience. We have memories, some factual, some farcical, that are part and parcel to the content of our dreams. Some of these memories can be introspective descriptions of mind, or at least what the border of mind appears to be.

MEMORY AND DREAMS

Consider the dream as a memory remembered from sleep.

Both memories and dreams are cognitive processes that require the accessing of information stored in the CNS. In a memory this information is transferred into the conscious thought of waking. In a dream this information is first integrated into the sleeping brain before being transferred to waking.

Many memories recall perceptual input received from the various senses. Perception effects memory, and memories can be altered at the level of perception. Dreaming does not generally incorporate perceptual input. Dreams are occurring during sleep, a state of minimal to non-existent perceptual input, and dreams cannot be easily altered by perception. Outside environmental noxious input such as loud noise, pain, and pressure is sometimes incorporated into dreaming in sleep stages that are closer to wake such as sleep onset and rapid eye movement (REM) sleep (Kuiken & Sikora, 1993). Although such events of perceptual incorporation into dreams are the exception rather than the rule, they have led to physiologic theories of dreaming which propose that one of the functions of dreams is to convey information about bodily functioning (Garfield, 1991). If we consider the dream as a memory remembered from sleep, this process of perceptual integration would occur as it does in the process of perceptual integration into memory. In memory, the neuroanatomic site at which this process occurs is at the level of the cortex and the sensory neculi of the brain stem (Tulving, 1983).

Another point at which the cognitive process of memory can be affected or altered is at the level at which the perceptual information of the memory is integrated with stored information. The perceptual "chair" that we are seeing in front of us is integrated with the perceptual memories of "chair." This process is called encoding or ecphory. Such encoding must then be transferred to conscious awareness. Both encoding and memory transfer are processes that occur in the hippocampus and cortex.

Once encoded the memory can then be organized and expressed. These processes occur in the massive area of the human brain called the frontal cortex. Such frontal-cortex-based processes are affected by different variables than perceptual input and encoding (Eich, 1989; Moscovitch, 1989).

IMAGERY AND INTRINSIC MEMORY

Dreaming is likely to utilize the same neural processing system as memory in its integration into conscious thought.

Dream differs from most memory in having only minimal perceptual (sensory) input, and no encoding required for that perceptual input. There are types of memories that have the same characteristics. One of these types of non-perceptually based memories is called intrinsic memory, which includes our memory of basic concepts. When we look at a form and describe it as a chair, that memory is rarely based on perceptual input to the particular chair. It is based on an intrinsic concept of "chairness" – the attributes which apply to many objects viewed from many differently appearing perspectives, which are all "chairs" (Roediger et al., 1989) (Fig. 9.1).

FIGURE 9.1 An intrinsic memory of "chairness."

Imagery is a type of memory process involved in the creation of new images that is also not based on perceptual input. Imagery can be considered our ability to develop "mental images" of objects without actual perceptual input. Imagery can be addressed at different levels, including those of image generation, image inspection, image maintenance, and image transformation (Kosslyn, 1994). Imagery has much in common with dreaming besides its perceptual dissociation. It is often visually based and replete with multiple associations (Marks, 1990).

Because imagery and intrinsic memory do not include perceptual input, they cannot be experimentally studied with the same techniques used to study learning and perceptual-based memories. We can, however, study the transfer of imagery and intrinsic memory into conscious awareness, and have developed techniques for experimentally evaluating the organization and expression of these non-perceptual types of memory. It has been proposed that dreaming utilizes the same neural processing system as memory in its integration into conscious thought (McCarley & Hobson, 1975). If this is true, we should be able to study dreaming as we study imagery and intrinsic memory – at the levels of transfer to conscious thought (hippocampal/cortex) and levels of organization and expression (frontal cortex).

In order for a dream to influence conscious thought, the dream must first be remembered or recalled while we are awake. Once we recall a dream, we develop associations both intrinsic and extrinsic to that image and often try to address the meaning of the dream. At this point, many of us will share the dream either verbally with another individual or write the dream down in a journal. It is at this point the dream becomes available for outside input, interpretation, and study (Cavallero & Foulkes, 1993).

DREAM RECALL

We can study whether a dream is remembered on awakening – dream recall. Dream recall occurs from all stages of sleep with recall affected by the stage of sleep from which the dreamer is awakened (Foulkes, 1985). Dreams are reported at highest levels on awakenings from sleep onset (stages 1 and 2) and REM sleep (>80%). Recall from other non-REM stages of sleep is lowest for deep sleep from which only 40% of awakenings are associated with dream recall. It has been postulated that this difference in sleep stage recall is dependent on the sleep stage of the dreams' distance from waking. If the dream occurs in a sleep stage closer to waking or if the individual wakes from the dream (both processes typical of REM sleep dreaming and sleep onset dreaming), there is a greater chance that dream will be remembered (Koukkou & Lehmann, 1983).

Dream recall declines rapidly with increasing time after the REM period, and increasing time after waking (Goodenough et al., 1965). If you wake an individual gradually rather quickly, dream recall declines (Goodenough, 1991). Waking during or immediately after dreaming may be required for consolidation of the memory trace so that the dream is remembered [29]. Koukkou and Lehmann (1983) have incorporated this sleep state dependence of dream recall into an "arousal–retrieval" model. They propose that short-term dream recall declines rapidly based on the time since the dream occurred and the time since arousal from sleep memory trace of the dream. Evidently the dream trace is available for only a short time (seconds or minutes) for incorporation into the long-term memory store during waking (Goodenough et al., 1965).

It has been suggested that dream recall can be affected by the brain's slow transition from sleep to wakefulness. Waking is comprised of not just one state, but a series of states with at least as much variability as the various stages of sleep (Hartmann, 1998). Sleep disorders such as obstructive sleep apnea that typically result in daytime sleepiness are also associated with a decline in reported dream recall frequency (Pagel & Vann, 1995). Individuals that are more likely to fall asleep during wakefulness report lower levels of dream recall (Myers & Pagel, 2001).

Another variable affecting dream recall is dream content. Recall is affected by dream intensity, with the most intense or "transforming" dream more likely to be remembered (Cohen, 1979). Dream salience – the greater novelty, bizarreness, affectiveness, or intensity of an experience – increases the potential for recall of a particular dream. The REMS dream remembered in the morning is most likely to be from the longest and most physiologically disturbed REM sleep period (most eye movements and irregularity of respiration) (Goodenough, 1991). Some sleep disorders induce an increase in REMS pressure, so that when the individual does fall asleep, the individual is likely to enter directly into prolonged episodes of REM sleep. These sleep disorders include narcolepsy, posttraumatic stress disorder (PTSD), and some of the mood disorders. When patients affected by these disorders fall asleep, they transition almost immediately into REM sleep rather than taking the normal 60–90 minutes until the first REMS period. Such patients have been reported to have increases in both dream and nightmare recall when compared to a similarly somnolent population (Myers & Pagel, 2001).

Dream recall can be affected by underlying illness, decreasing in patients experiencing depression and obstructive sleep apnea (Armitage, 1994; Pagel & Vann, 1995). Dream recall can be lost after some forms of brain trauma (Solms, 1997). Younger women are more likely to remember dreams than older men. Gender differences exist in dream recall. Female recall is consistently higher than male recall in almost all population studies that have addressed dream recall frequency (Gimbra, 1979; Pagel & Vann, 1992). Some researchers have described increased dream recall in individuals with interpersonal attachment styles classified as "insecure attachment" and "preoccupied" and in individuals prone to specific types of visual dream imagery (Wolcott & Strapp, 2002). Most current studies suggest that personality and ethnicity have minimal effects on levels of reported dream recall (Blagrove, 2000; Pagel & Vann, 1993). However, if an individual has a higher level of interest in the dream process, he or she is more likely to recall dreams (Strauch & Meier, 1996).

DREAM CONTENT

We remember dreams not as real experience but as illusions that strike us on waking as having been dreamed (States, 1997). We can study reports of the thought, memory, imagery, and associative content of these illusions. This has been the focus of dream studies for millennia. The study of dream content involves at least two memory processes, including the associations of this information with other stored memories (ecphory) and the organization of this material for expression. Content studies require that the dream be written or described to a third party. In situations in which the dream is described to another person, the interaction between the dreamer and the transcriber of the dream are known to affect the described report. This interaction can result in psychoanalytically described repression and transference affecting the transcription of dream content (Ekstrand, 1972; Reyna, 1995).

Dream interpretation has a long and convoluted history, with attempts at interpreting dreams characterizing all human societies. There is, of course, not just one way to interpret dreams. Each social group develops its own technique of dream interpretation, sometimes similar or shared, sometimes dissimilar from others (Tedlock, 1992). This variety of approaches is not limited to social groups, for it seems that almost every researcher of dream content has developed a new scale for analysis of dream content based on that researcher's area of interest (Kramer & Roth, 1979). Because of the many potential interpretations that can be applied to dreams, modern psychologists have attempted to develop more scientific and replicable approaches to the study of dream content. Dream length and types of content have often been the focus of study. This analysis most commonly involves use of the Hall and Van de Castle scoring method that was developed in the 1960s to analyze the words and their meanings included in a dream report (Hall & Van de Castle, 1966). Researchers use the Hall and Van de Castle scale not because it is specifically best for a particular study, but because of its extensive history of use and validity on analysis. In an attempt to minimize the transference and effects of researcher bias effects that bedevil many of the attempts to study dream content, this approach has recently been computerized for use in comparative dream analysis of large dream series (Domhoff, 1998). But there are still methodological problems with dream content studies. These new approaches to studying dream content may yet support years of psychoanalytic research suggesting that dream content can be altered by personality, gender, social background and ethnicity, concomitant psychiatric illness, stress and trauma, and by daily life events (Kramer, 1993; Domhoff, 2003).

EVERYDAY MEMORY AND MIND

Our personal collection of memories are incorporated into the content of our dreams. Most often these are recent memories of daily experience. This reflects the "continuity hypothesis" that dream content has its highest correlation with an individual's life and waking experience (Domhoff, 2003). In dreams such "day-residue" can interact with stored long-term memories. Some of those memories may be bizarre or incorrect approximations of past experience. This has led DeWitt (1988) to suggest that in dreams the normal waking memory processes of reality construction and reality monitoring may be impaired or inoperative. DeWitt evaluated the customary understanding his psychology students had of the memories present in their dreams. The members of his classes considered the following attributes true of dreaming (the percentage of his students holding a particular view is listed in parentheses):

1. Visual images are vague and unstable (46%).
2. Factual information is "just known" without any evidence (56%).
3. Bizarre occurrences are accepted as normal (67%).
4. Events are grossly misinterpreted (46%).
5. There are abrupt shifts of continuity and scene changes (70%).
6. The dreamer assumes different identities (59%).
7. The dreamer has multiple points of view often inconsistent with each other (62%).
8. Prior events within the dream are invented (46%).
9. Earlier parts of the dream are distorted (36%).
10. Events that are anticipated or feared turn into actual happenings in the dream (60%).

These findings indicate that many of DeWitt's psychology students believe that during dreaming, normal processes of reality construction and reality monitoring are impaired or inoperative.

Dream memories are internal memories that differ from the waking memories of external perceptual experience. Waking memories make sense in terms of our knowledge of the world. A memory from dream is more likely to violate natural laws or conflict with other knowledge. We can fly in dreams and confront enemies in socially unacceptable ways. External memories are richer in sensory attributes such as sound, color, and texture, are more detailed and are set in the context of time and place of occurrence and other ongoing events (Cohen, 1996). Dream memories have less sensory and contextual detail, but often include cognitive operations such as reasoning, inferring, and imaging.

In dream memories, the dream content differs from our waking memory of perceptual experience. Dream memories are not repressed. Dream memories can be merciless, excluding nothing that may lie in their mnemonic path. Dream memories are images altered by the passage of time since the perceptual experience. Eventually these memories are incorporated into a dream that can have a maintained state of present experience for the dreamer that is maintained as long as the dream is remembered (States, 1993). Images from memory can be experienced in dream independently of waking consciousness and returned to memory without exposure to waking thought. It is little wonder that such memories are often used by psychoanalysts to explore aspects of the mind.

DREAM USE

Dreaming can also be studied at the level of dream use or incorporation into behavior (Spear, 1994). Dream use has been shown to be higher in females, successful filmmakers, and in individuals with higher frequencies of dream recall. Increasing age and obstructive sleep apnea are negatively associated with the use of dreams in waking behaviors. The environmental stress of natural disaster (Hurricane Iniki) did not affect levels of dream use. Racial background and ethnicity have minimal effects on dream incorporation into behavior (Pagel & Vann, 1992, 1993, 1995).

The incorporation of memory into behavior is a frontal cortex process. In individuals with behavioral abnormalities resulting from frontal cortex injury or dysfunction, consistent deficiencies in neuropsychological function are known to occur. Frontal lobe cognitive functions are generally described as executive functions. They include planning, organization, execution of complex goal-directed behavior, accurate self-awareness, flexible response to changing environmental contingencies, persistence in a task or maintenance of response despite distraction, and creative problem solving (Duffy, 1994; Fogel, 1994). Both executive functions and dream use are altered by illnesses such as obstructive sleep apnea and mood disorders that affect frontal cortex functioning. Considering the dream as a memory from sleep, the incorporation of dreaming like the incorporation of memory into waking behavior is most likely to be a frontal-cortex-based executive function (Pagel & Vann, 1996).

Current research derived from studies of dream that use cognitive models of memory is summarized in Table 9.1. Different variables affect dreaming at the different levels of the incorporation of dreaming into waking cognition. Dream recall is affected more by sleep associated variables, dream content by personality and personal interest variables, and dream incorporation into waking behavior by social interest variables.

TABLE 9.1 Variables Affecting Dreaming at Different Levels of Incorporation into Waking Cognition			
Variables (Levels of Effect)	Dream Recall	Dream Content	Dream Use (Incorporation into Waking Behaviors)
Sleep stage	$(+++)$	$(0, ?)$	(ns)
Time since waking	$(---)$	$(-, ?)$	(ns)
Daytime sleepiness	$(--)$	(ns)	$(-, ?)$
Gender	$(+++)$	$(+++)$	$(+++)$
Age	$(--)$	$(-)$	$(-)$
Salience	$(++)$	$(++)$	$(++)$
Ethnicity	(0)	$(++)$	(0)
Personality	(0)	$(+++)$	(0)
Creative interest	$(+)$	$(++)$	$(+++)$
Stress – life events	$(+)$	$(+++)$	$(+)$
Illness			
OSA	$(--)$	$(?)$	$(-)$
Depression	$(--)$	$(-)$	$(-)$
PTSD	$(+)$	$(++)$	$(++)$

Note: (ns): not studied; (?): conflicting studies; (0): no effect; (+,−): mild positive/negative effect; (++,− −): moderate positive/negative effect; (+++,− − −): large positive/negative effect.

MEMORY AND MIND

One realization that becomes obvious from addressing dreaming as memory is that the process of memory is very much about the correct maintenance of our waking relationship to the external world. Memory during waking is utilized in the maintenance of the consistent structure of learned information and repeatable patterns of perceptual experience. We receive schooling to expand our capacity to correlate waking memory with our memory performance that is judged based on the degree that our memory equates with real-world experience. This approach discounts any experience that cannot be shown to have concrete perceptual correlates. Dream memories often fail such tests of reality testing. The rules of the dream world often differ from those of perceptual reality. People fly, animals talk, and events are portrayed that have not actually occurred. Dreams confound and associate memories that are likely not to have occurred together in the external world. Taken to the extreme, such a perspective can result in dreams being viewed as impaired, potentially worthless cognition needing to be cleared from the neural net during sleep.

Memory gives us a logical time-based framework through which to comprehend our external world. But what, if as Descartes suggested so long ago, mind includes those cognitive processes that we cannot prove or see even with technical extensions of our perceptual expertise. Then those illogical false memories, those irrational, invented misinterpretations of dream, are the memories with the potential to tell us about the internal world of our minds. And then the reality based cognitive modeling of waking memory, our process of organizing experience, learning, and putting perceptions into storage, that can stand in the way of our comprehending mind, blocking our ability to consider whether something else – mind itself – even exists.

CHAPTER 10

The Brain/Computer Interface

The most sublime intelligence will never be able to find in a closet, what only exists in the vast field of nature.

(Franz Joseph Gall, 1938, Vol. V. p. 317)

INTELLIGENCE

Intelligence is also a difficult cognitive term to define. In 1575, the Spanish physician Juan Huarte defined intelligence as the ability to learn, to exercise judgment, and to be imaginative (Russel, 1927). Intelligence can be operationally defined, consisting of specifying a goal, assessing the current situation to see how it differs from that goal, and applying a set of operations that reduce the differences (Newell & Simon, 1972). Intelligence is often equated with purpose and complexity, with "intelligence" corresponding to both a broad range of abilities and the efficiency with which they are accomplished (Calvin, 1996). Intelligence also implies flexibility and creativity, an "ability to slip the bonds of instinct and generate novel solutions to problems" (Gould & Gould, 1994, p. 149). Despite such a long history of varingly applied definitions, intelligence is often operationally defined as a level of educational or cognitive capacity based on performance on standardized IQ tests.

ARTIFICIAL INTELLIGENCE

A major accomplishment of the last half of the 20th century was the progress made in the development of artificial systems that can serve as extenders or facilitators for human intelligence. Current systems have both perceptual and computational capabilities of amazing complexity. Systems have been designed with functional capabilities that include perceptual, computation, reasoning, communication, and motor activities that are beyond human capability. Artificial intelligence (AI) systems can respond to changing outside environmental information without additional human input. Based on many of our definitions, artificial systems can now generally be described as intelligent, independent of their roles in facilitating human intelligence.

Science fiction authors have suggested that AI marks the next stage of planetary evolution (Fig. 10.1). As we discussed in Chapter 7, it has been proposed that consciousness may be an emergent phenomena of complexity (Davies, 2006). Crick and Mitchinson (1983, 1995) have proposed that aspects of mind may develop as a natural result of neural net overloading. Perhaps we are on the verge of the next evolutionary stage. Complex artificial systems have been created that could potentially become conscious and have aspects of mind. But AI systems have not yet been shown to have development of aspects of mind. We utilize our understanding of cognitive systems to create and manipulate AI systems. By pushing to the edge of current capabilities and delineating what artificially intelligent systems can and cannot accomplish, we can better define what lies beyond the capacity of such artificially created systems, beyond the limits of our current capabilities to artificially parody neural function. Aspects of mind are not only those

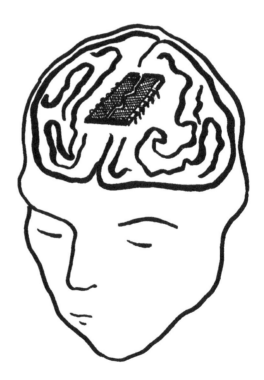

FIGURE 10.1 The next stage of planetary evolution?

which we cannot scientifically explain, they are also those which we cannot artificially create. Exploring the limits of the capabilities of artificially intelligent AI is another approach to discovering what lies beyond the capabilities of these systems – what we are calling mind.

The Perceptual Interface

A hardwired neural interface between the human brain and accessory computer intelligence is a common, accepted feature of futurist literature, films, and graphic novels. In some ways, we are rapidly moving closer to that reality. Newer AI systems utilize knowledge of biologic sensory systems in their motion, tactile and propioceptive motor sensors. The computer interface is attempting to parody our perceptual capabilities as well as the central nervous system (CNS) coding systems used in the process of memory integration that we utilize in understanding our external environment. The perceptual capabilities of artificial systems are far in excess of human capacity, utilizing lenses to see at distance, and sensors to assess and utilize electromagnetic wavelengths beyond the biologic range.

Virtual reality is conceptually possible, and already present on a very real level in military combat. Weapons are aimed and soldiers are trained virtually on war game computer systems that utilize a complex visual, vocal, tactile, and auditory interactive interface. We have reached the point where the experience of warfare includes death on screen – a final concrete conflation of computerized representation with a terminal reality.

The Motor Interface

AI systems can be used to extend our perceptual and motor systems beyond body control to control of our external environment. Robotic systems have been developed with perceptual, motor, and analytic capabilities far beyond human capabilities. The Mars Landers are examples of current system capabilities. They have traveled to and explored complex environments biologically toxic to humans. In this environment they have generated their own energy, traveled hundreds of miles, handled navigational tasks, completed complex astronomical, climatic and geological experiments and broadcast that data back to human handlers on another planet. Systems have also been developed capable of producing emotional responses in humans. Toys such as "Tickle-me Elmo" are prominent examples of the emotional potential of artificial systems.

Computation and Thought

Computers have fabulous memory systems. The existence of calculators and computational systems with skills far beyond human capacity has led to questions as to the definition of intelligence. If the process of thought is defined simply as the manipulation of symbols, computers excel at the task. If the process of thought is more strictly defined as the ability to initiate original thoughts, the capabilities of computers to think become questionable. The capabilities of AI systems have led to the general acceptance of a definition of intelligence conveniently amenable to human capabilities: intelligence consists of specifying a goal, assessing the current situation to see how it differs from that goal, and applying a set of operations that reduce the differences (Newell & Simon, 1972). The human brain represents the external environment visually, generally describing the environment with a phonological speech representation that is couched with a grammar of words arranged in hierarchical trees. Those representations can then be approached conceptually (Pinker, 1994). In order to better interact with humans, AI systems have been developed with many of these same capabilities. AI systems have visual recognition capacity and the ability to phonetically understand human speech. Such information can be grammatically represented and classified before conceptual manipulation. Although this seems a convoluted approach to cognitive processing, it is a biologically based approach that can be accomplished by artificial systems.

The vast majority of current computer systems are serial processors operating based on a series of on–off switches. Such systems are excellent at computation and memory storage. The human CNS does not operate serially, or depend specifically on on–off switch transmission of information. The CNS operates on multiple, parallel, and simultaneous levels (Chapters 4–7). Biological systems can be modeled at different levels of resolution using different techniques for analysis at each level. For example, the information processing of a single neuron can be studied at the level of chemical reactions, the flow of those chemicals across the cell membrane, the ionic flux, electrical field effects of such changes across the membrane, and as a sequence of states potentially resulting in a spike potential. Neural information processing can be modeled further at the level of the synapse, the neuron, the neural network, and the brain as a whole. Each of these systems has a characteristic methodology of operation that operates both independently and in sequence with the other processes. The end result is a vastly complex biologic parallel operating system (Sampson, 1984).

The operative thought processes characteristic of serial AI systems are much different than the types of complex parallel processing involved in the human thought (Benson, 1994, p. 14). Basically simple process of serial on–off switches achieve processing capability based on the quantity of potential connections operating at high speed. The achievement of an answer to a mathematical question by a serial computer is not based on a limited number of varied logical steps, but on a repetitive application of the same logical steps repeated in consideration of all possible alterative answers to that mathematical question. There are current attempts in process to create AI utilizing different types of parallel processing capacities similar to biological systems. Neural net systems (Chapter 7) utilizing massively parallel processing have found commercial applications in digital computation systems addressing computational problems that are poorly amenable to the linear mathematical approaches utilized by serial processing systems. Future AI systems will be developed based on parallel processing computational systems in order to parody human operative thought processing. Such systems could potentially facilitate human–machine interaction. It is possible that parallel processing AI systems based on neural nets will "think" in a manner that is more like the human. But it should be remembered that these vastly complex neural nets model the complexity of the biologic information processing system only at the level of interactive, computational on–off switches. As noted in Chapters 4–6 each biologic neuron has the capacity of responding to a multiplicity of inputs, multiple neurotransmitters, neural–hormonal effects, and external electrical fields. Each level of additional alternative neural input can logarithmically expand additional levels of complexity to the operating of these systems.

Communication

The Mars Landers communicate using on–off bits (e.g.: 101100100). For such a one-dimensional system, the complexity of information is based on the quantity of such on–off bits required for the encoding of a message (Pierce, 1980). This telemetry data can be broadcast interplanetary distances at high speeds. A system with the capacity to transmit 1000 binary digits per second is transmitting 1000 information bits each second. At the receiving station on Earth, computer programs transfer that bit data into numbers and words that constitute a computer language. The programmer on Earth communicates back to the Lander giving orders, sometimes suggestions, in the same way. Interactive translation systems to verbal speech are in general use in automobile navigation systems and hospital dictation systems as well. These systems have the capacity to translate into multiple phonetic and computer-based languages. The capabilities of such systems in modes of translation are already beyond human capacity. However, translation capabilities are currently at the level of parroting or direct translation of text. The next step will be to teach artificial systems to develop the vocabulary and performance communication skills required by a true language. Once a machine has language, and the process of interaction develops beyond just communication, the possibility exists for the machine to develop an individual personality (Lenhert & Ringle, 1982).

Problem Solving

AI can be used to solve a wide variety of problems using currently available search, reasoning, and knowledge-based systems (Caudill, 1992).

Probability and fuzzy reasoning approaches allow current systems to deal with uncertainty in the information and processing rules that define data stored in large knowledge bases (Pearl, 1988). Brain-like neural networks recreate the massively parallel neuron-to-neuron line interactions of the CNS. In the future, it is likely that capacities of artificial systems in problem solving will become even more within the realm of biological capacity (Penrose, 1990).

The Capabilities of Artificial Systems

The capabilities of artificial system are a useful marker of cognitive processes that we scientifically understand and can recreate using our technology. Current systems have no sense of self-awareness, no recognizable personality, and no obvious aspects of mind. The creation of artificially intelligent systems has yet to take us any further down the road toward making contact with the mind itself. However, specialists in AI tend to believe that mind-based capabilities such as creativity may be an emergent by-product of increasingly complex artificially intelligent systems. Some AI researchers have proposed that when a system becomes sufficiently complex and intelligent, self-awareness and mind can be expected naturally to follow (Davies, 2006). Such capabilities would not be programmed but would develop based on interactions between programmed subsystems. It is possible that we may understand these interactions no better than we understand the operation and functioning of our own minds, however having a belief in the consciousness of extremely complex systems leads to interesting postulates (Caudill, 1992). If extremely complex systems are conscious, the cosmos on an extra-galactic scale must surely be a conscious system. Earth itself can be viewed as a massively parallel processing system of multiple complex components. Based on such a "Gaia" hypothesis, it is conceivable that the planet Earth is a conscious system. Our interconnected electrical transmission line systems when hooked to such complex outputs as digital televisions and personal computers achieve an extreme level of complexity. Theoretically, this system should easily have the complexity to initiate consciousness. That is an interesting possibility to consider the next time that your lights flicker.

Descartes was but the first of many that have found it useful to talk of the brain as a machine (Pogliano, 1991). Analyzing current AI system capacity in perceptual and motor interfacing, computation, communication, and problem solving gives us an excellent marker of the cognitive capabilities that are likely in the human to be brain rather than mind based. We understand these cognitive processing systems well enough to artificially create systems with specific capacities far beyond those of biological systems.

Postulating that machines can become conscious remains a staple of science fiction. Some scientists have developed hypothetical theories where dreams could potentially have a function even for serial processing systems. The processing complexity of artificial systems is limited when compared to biologic systems, and most of these individuals are theorizing outside their field of specialty. Such simple hypotheses do suggest possibilities of neural functions for dreaming, functions that are likely even more complex in the biologic reality of multiple parallel information processing systems in which dreams are but one of the cognitive processes taking place. Currently, there is little or no proof that artificial systems have developed dreaming. Currently, there is little evidence that artificial systems have cognitive capabilities that reflect the development of mind.

SECTION 4

The Cognitive Process of Dreaming

Vision, thought, and emotion are cognitive processes that occupy much of our waking lives. The basic components of dream content include visual images, memories, thoughts, and emotions. It is likely that in dreaming we utilize the same brain-based neural processing systems for these cognitive processes that we use in waking. We understand much of the brain basis, the neuroanatomy, the neurochemistry, and the electrophysiology, of some of these cognitive processes involved in dreaming. Other processes, we understand less well.

We are visual creatures. Damage to visual perception systems is immediately obvious and has been the focus of medical research. Large portions of the central nervous system (CNS) are involved in vision, and multiple animal models are available for analysis. Because of this work, we understand the brain-based substrate of vision better than almost any other set of CNS systems. The content of dreaming is primarily visual. Based on current technology, the visual component of dreaming is a process that is within our capacity to artificially create.

We also understand something of the neuroanatomy and neurochemistry involved in the cognitive processing of emotions. Emotions have evolutionary survival value utilized in the propagation of the species with emotional expression documented in many species besides humans. There are known brain-based CNS correlates of emotional processing. The emotions that figure so prominently in dreaming are likely to utilize the same systems involved in the processing of emotions when awake. One of the primary functions proposed for dreaming is as an emotional regulatory system. Disorders of emotional processing can result in disordered dreaming. Such diagnoses are amenable to modern medical approaches to diagnosis and treatment with some of the approaches used in therapy directly addressing and treating disease-associated disordered dreaming.

What our brains do is think. And it is difficult for most of us to stop thinking when we are awake. The thinking that occurs in dreams is both similar and different from waking thought. The brain-based CNS correlates for the processes of thought are poorly described with our current technical capacities. In approaching the process of thought we will again find ourselves reaching the borders of scientific knowledge. Aspects of thought appear to be mind based rather than brain based.

The emotional, memory, and visual components of dreaming are primarily brain based and amenable to modern scientific study. We can use the knowledge and techniques of modern neuroscience to understand much of these components of the dreaming. The emotion and visual imagery components of dreaming are brain based and amenable to currently available techniques of scientific study. It is only at the limits of these areas that we approach components of each process that approximate aspects of mind – the global feelings associated with emotions, and the hallucinatory imagery of the visual system. However, the cognitive process of thought, particularly the associative thought characteristic of dreaming and creative process, is poorly described by modern neuroscience. When we look at the cognitive processes involved in dreaming, it is through study of the intangible and invisible process of thinking that we most quickly approach the borders of what we do not understand. In the story and content of dream we find aspects of mind.

CHAPTER 11

Emotions and Dream

Is it not monstrous that this player here,
But in a fiction, in a dream of passion,
Could force his soul so to his own conceit,
That, from her working, all his visage wann'd;
Tears in his eyes, distraction in 's aspect,
A broken voice, and his whole function suiting
With forms to his conceit? And all for nothing!

(W. Shakespeare, Hamlet, ACT II. Sc. 2)

The expression of emotion is a brain-based cognitive process. Emotions have well-described biological correlates. The study of the biological basis of emotion dates back to 1872 and Darwin's last great book *The Expression of Emotions in Man and Animals*. In this writing Darwin proposed an evolutionary tree of animal species based on the development of emotional expression from base organisms through intermediary species with the achievement of the highest expression of emotion in humans (1872). Emotions are considered to be a complex collection of neural responses forming a distinct pattern that are produced as a response to an emotionally competent stimulus (Damasio, 2001).

Specific emotions can be altered by discrete areas of damage to the brain with different brain areas controlling different emotions (Damasio, 2003). In the amygdale, visual and auditory emotionally competent stimuli trigger emotions of fear and anger (Amaral, 2002). Emotional triggering also occurs in portions of the ventromedial prefrontal cortex, and area of the frontal supplementary motor area and the cingulate. None of these triggering sites alone produces emotion. The expression of emotion requires subsequent activity in other areas of the central nervous system (CNS) including the basal forebrain, hypothalamus, and brain stem nuclei among others (Fig. 11.1) (adapted from Damasio, 2003). Damasio has suggested that we should define emotion as the external expression of internal feelings.

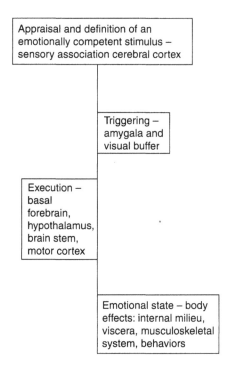

FIGURE 11.1 A diagram of the main stages of the triggering and execution of emotions and associated brain processing areas.

FRIGHTENING DREAMS

Negative emotions are a part of many dreams. In some dreams emotion is all that is expressed or remembered on waking. This is particularly true of frightening dreams such as nightmares and night terrors. Extreme emotions characterize the arousal disorder known as the night terror. Night terrors are one of the arousal disorder parasomnias that include sleepwalking (somnambulism) and confusional arousals. These arousal disorders occur during partial or complete arousal from stages 3 and 4 of deep sleep. Night terrors and confusional arousals are most common in children, characterized by confusion and autonomic behavior (sweating, flushing, and dilated pupils) on waking, difficulty waking from the event; fragmented imagery of dream content, rapid return to sleep, and amnesia for the event the next morning. Sleepwalking can occur with these types of frightening dreams and include surprisingly complex behaviors. The child can be seen negotiating obstacles and carrying out seemingly purposeful tasks, as well as inappropriate behaviors, such as urination. During confusional arousals, the child may seem to be awake yet considerably confused. The child may cry, yell, moan, or speak in unintelligible sentences and may or may not recognize the parental figure. A "blood-curdling" scream, sweating, flushing, and dilated pupils accompany the most extreme form of confusional arousal, the sleep terror. Sleep terrors often frighten the parent even more than the child. Despite such an extreme presentation, the childhood arousal disorders are usually not associated with psychopathology. These sleep-related behaviors are experiences over which the sleeper has no conscious deliberate control (Table 11.1). Night terrors, what some consider the most frightening and emotionally powerful of dreams, occur outside of rapid eye movement (REM) sleep.

Nightmares differ significantly from night terrors (Table 11.2). Nightmares have elaborate mental content, while the mental content of night terrors is usually sparse or absent. Night terrors are more likely than nightmares to be associated with vocalizations, somnambulism, and autonomic discharge. Night terrors are more likely to occur early in the night (1–3 a.m.) during the first deep sleep period while nightmares are often associated with the last and longest REM period of the night. Individuals experiencing night terrors are difficult to arouse, while those experiencing nightmares often awaken from the dream.

Typically, nightmares are coherent dream sequences that seem real and become increasingly more disturbing as they unfold. The emotions in nightmares usually

TABLE 11.1 Parasomnias

Disorders of arousal (from NREM (stages 3 and 4) sleep)

1. Confusional arousal
2. Sleepwalking
3. Sleep terrors

Parasomnias usually associated with REM sleep

1. REM sleep behavioral disorder
2. Recurrent isolated sleep paralysis
3. Nightmare disorder

Other parasomnias

1. Sleep-related dissociative disorders
2. Sleep enuresis
3. Sleep-related groaning
4. Exploding head syndrome
5. Sleep-related hallucinations
6. Sleep-related eating disorder

TABLE 11.2 Nightmares and Night Terrors: Distinguishing Characteristics

Night Terror	Nightmare
Associated with arousals from deep sleep (stages 3 and 4)	Associated with REM sleep
Intense: vocalizations (blood-curdling scream), fright, somnambulism, autonomic discharge	Intense: vocalizations, fright, motility, autonomic discharge
Sparse mental content – amnesia	Elaborate mental content – less amnesia
Difficulty in arousing individual	Often associated with arousals from sleep
More likely to occur early in the night	More likely to occur late in the night
In childhood (2–4%)	Most common in children (40–50%)

involve anxiety, fear, or terror and frequently include negative emotions such as anger, rage, embarrassment, and disgust. Dream content most often focuses on the possibility of imminent physical danger to the individual (e.g., threat of attack, falling, injury, death) but may also involve aggression toward others, potential personal failures, suffocation, and other distressing themes. On awakening, individuals can usually give a detailed description of the nightmare's contents. Multiple nightmares within a single sleep episode may occur and may bear similar themes.

Some personality patterns are typically present in individuals with frequent nightmares. These personality characteristics include fantasy proneness, psychological absorption, dysphoric daydreaming, and "thin" boundaries (Hartmann, 1994). Such individuals are more likely to have a creative or artistic focus in their daily lives. Some of these individuals may utilize their dream and nightmares in highly successful creative careers in writing, acting, and film (Pagel et al., 1999). Individuals with frequent nightmares are prone to psychopathologies including the schizophrenia-spectrum disorders, anxiety disorders, and dissociative disorders. An individual's level of nightmare *distress* is much more robustly associated with psychopathology than is nightmare *frequency*. Associations with psychopathology have been described in adults and adolescents.

Up to 50% of children have frequent nightmares that are often drawn in description as monsters (Fig. 11.2). In children nightmares are generally not associated with psychiatric disorders, except for the specific association of nightmares occurring after major psychological or physical trauma in children with posttraumatic stress disorder (PTSD).

FIGURE 11.2 In younger children nightmares are most often described as the visual imagery of "monsters."

Nightmares are the most common symptom of PTSD, an anxiety disorder characterized by symptoms of "hyper-arousal" occurring after serious physical or mental trauma. Nightmares beginning within 3 months of a trauma are present in up to 80% of PTSD patients. Nightmares present after a trauma can predict the delayed onset of PTSD. PTSD nightmares are more likely to occur in females, individuals with low socioeconomic and educational levels, and those with prior psychopathology. The severity, duration, and proximity of a traumatic event are the most important risk factors for PTSD. Social support, family history, childhood experiences (including previous trauma), personality variables, and pre-existing mental disorders may affect the development of PTSD. However, the disorder can develop in individuals without clear predisposing conditions, particularly if the stressor is extreme.

Although approximately 1/2 of PTSD cases resolve within 3 months, posttraumatic nightmares can persist throughout life. Nightmares arising either immediately following a trauma (acute stress disorder, ASD or 1 month or more after a trauma (PTSD) can occur during NREM sleep, especially stage 2, as well as during REM sleep and at sleep onset (Francis et al., 1994). Posttraumatic nightmares often take the form of a realistic reliving of a traumatic event. Some PTSD nightmares may depict only some of the elements of the traumatic experience in realistic or symbolic form. ASD and PTSD nightmares can develop at any age after physical or emotional trauma. An individual with PTSD is at risk for severe depression, marital conflict and divorce, loss of job, self-destructive and impulsive behavior, dissociative symptoms, multiple somatic complaints, hostility, social withdrawal, substance abuse and nightmares after withdrawal from substance abuse, survivor guilt, despair, and hopelessness (Pagel & Nielsen, 2005). It is not known to what extent the nightmares experienced by patients with PTSD contribute to these negative outcomes.

EMOTIONS AND DREAMING

The dreams that we remember are often those charged with affect. The waking experiences that are most likely to be incorporated into dreaming are those with emotional content (Piccione et al., 1977). Affective and emotional behaviors are the waking behaviors most likely to be affected by dreams (Ekstrand et al., 1977). Some studies suggest that an individual's mood before sleep changes during a night of sleep (Kramer, 1993). After a night of sleep most individuals are less unhappy and less unfriendly than they were at sleep onset. This finding has led to theories that dreams have a mood regulatory function. Dreams that reflect the waking emotional experiences of the dreamer, change across the night. Such emotionally charged dreams are linked to the emotional preoccupations of the dreamer the next morning. Successful dreams, based on this theory, are those that utilize progressive figurative problem solving during the night to resolve emotional conflict. In the morning after a night of successful dreaming, the dreamer has a decrease in emotional conflict and an increase in happiness (Kramer, 1993).

Nightmares and frightening dreams may reflect a failure in this mood regulatory system. Freud described failures in this system for subjects unable to ignore or assimilate experiences of profound unpleasure associated with trauma:

> We describe as "traumatic" any excitations from outside which are powerful enough to break through the protective shield. It seems to me that the concept of trauma necessarily implies a

connection of this kind with a breach in an otherwise efficacious barrier against stimuli. Such an event as an external trauma is bound to provoke a disturbance on a large scale in the functioning of an organism's energy and to set in motion every defensive measure ... There is no longer any possibility of preventing the mental apparatus from being flooded with large amount of stimulus, and another problem arises instead – the problem of mastering the amounts of stimulus which have broken in and of binding them, in a psychical sense, so that they can be disposed of (Freud 1916/1951).

Individuals with PTSD have experienced a physical or psychological trauma that is too extreme for the individual to emotionally integrate. When such an emotionally competent stimuli cannot be emotionally processed by this system, the individual develops recurrent frightening nightmares of the experience as he or she relives the experience during sleep. PTSD associated nightmares are often associated with increasingly negative moods on arousal marking a repeated failure of emotional integration of the negative experience (Levin & Nielsen, 2006). Mood regulation appears likely to a primary function of the dreaming process.

This theory fits well with the diagnostic profile of disorders associated with nightmares. Nightmare disorder occurs in individuals that have recurrent episodes of awakenings from sleep with recall of intensely disturbing dream mentation, usually involving fear or anxiety but also anger, sadness, disgust, and other dysphoric emotions (nightmares). Individuals with nightmare disorder do not have the history of a major psychological or physical trauma that is present in individuals having nightmares associated with PTSD. Individuals with nightmare disorder often have less nightmare distress than in those individuals with PTSD nightmares (Levin & Nielsen, 2006; Pagel & Nielsen, 2005).

EMOTIONS AND MIND

Emotions have evolutionary value, useful in assisting with the procreation of the species and through flight and fight, keeping the individual alive to pass on genetic material to the next generation. Dreams are full of emotional process, likely to serve a functional role in the emotional processing system. This emotional processing system, in large part biological and brain-based, functions in the cognitive process of dreaming.

While emotions and their expression are aligned with the body, feelings can be considered as a part of the "mind" component of the emotional process. Feelings, so defined, are thoughts with themes consistent with the emotion. Feelings constitute a mode of thinking, a style of metal processing, that integrates and addresses the emotion (Damasio, 2003). One way that feelings differ from emotions is that the perceptual objects and events at the origin of feelings are internal rather than external. Based on such a definition, the emotional processing that occurs in dreaming could be considered the process of "feeling" rather than emotion. Since dreams in sleep rarely contain perceptual input, dream emotions by this definition are feelings – thought and process about emotion.

Feelings apparently utilize the same neural processing system as emotions. In positron emission tomography (PET) scanning studies of individuals asked to concentrate on an emotional life experience, perceptual somatosensory cortex shows activation (Damasio et al., 2001). This finding indicates that at least some of the same perceptual systems that are activated in the process of emotion are activated in the process of feeling. Feelings can be considered as the expression of human flourishing or human distress. The change in mood or happiness that occurs after a night of sleep reflects a change in feeling. Damasio (2001) suggests that such feelings are based on an actual brain substrate that inscribes a map in brain neural nets of the emotional and perceptual input associated with the experience of a particular emotion. In this way feelings interact with the neural network maps of emotional processing produced by perceptions associated with emotional experience and inscribed into memory.

EMOTIONAL DREAMING

Emotions are a major part of dreaming. Dreams are likely a functional part of an emotional processing system required for the assimilation of negative life experiences. The emotional system of the CNS has an evolutionary and biological basis. Dysfunctions in this system can result in psychiatric disorders such as PTSD that include symptoms of disordered dreaming. The neurological emotional systems and neural interconnections of this emotional processing system can be mapped in the CNS with our modern technological capabilities.

Much of the emotional component of dreaming appears to be brain rather than mind based. Because of that brain basis, this major component of dreaming appears to be amenable to modern neuroscience research. There may be an actual site in the CNS where a neural net is inscribed that reflects the memory of an emotional experience. In the future, the technology and techniques of modern medicine are likely to be utilized in the diagnosis as well as the treatment of disorders affecting the emotions of dreaming. The association of biologically based emotions with thoughts and feelings about those emotions remains a manner of contention often dependent on definition. The association between emotion and feeling lies at the limits of our understanding of how emotions may be developed and saved as memories in the CNS.

CHAPTER 12

Exploring the Visual Interface

*If a man could pass through Paradise in a dream, and have a flower
presented to him as a pledge that his soul had really been there, and if
he found that flower in his hand when he awoke – Ay! – and what then?*

Samuel Taylor Coleridge (1912)

VISUAL CONSCIOUSNESS

Our perceptual systems are the coding systems that we use to understand our external environment. These systems present the external environment to our brains in a comprehensible manner. This is not a simple cognitive process. Vision, the most complex of perceptual systems, topographically requires the largest area of the central nervous system (CNS) of any cognitive function. This is not surprising. Vision is the primary perceptual system that we use in waking.

When we learn and represent spatial information we generally think in a visual modality (Klatsky & Lederman, 1993). Positron emission tomography (PET) scans of visual processing indicate that many varied areas of the CNS are required for vision. Multiple interactive CNS systems are involved in the perceptual process and cognitive evaluation of visual input. Techniques such as PET scan, neuropathology, and animal single neuron studies, we helped us to understand the cognitive and neuroscience of the many parallel visual processing systems in the brain. At least two major sophisticated cortical systems function to integrate visual content into memory and conscious activity (McCarthy, 1993). There is also a major sub-cortical system that can function at non-conscious levels to control eye movements and cross-modal spatial processes. It is through this system that our visual system sometimes operates without out conscious control.

As visually oriented creatures we require these multiple, extensive systems in the CNS to assess, translate, and store visual information. We take the perceptual information available to us from our visual system, and neurologically process that information into a representation of the external universe that we can interact with and manipulate on a consistent basis. Our perceptual representation of external reality is, however, but one of many possible representations of reality. In our biosphere there are organisms that operate successfully in much different sensory environments: the tactile jellyfish, the dog with extended aural and olfactory range, the bird with limited color vision and extended acuity, or the fish with electromagnetic sensitivity. Our three-dimensional concept of external reality is likely only one among many possible representations available for the conceptual description of space. There are added dimensions that can be shown mathematically to describe external space as well or better than our three-dimensional coding system. Potentially a representation of external reality could include those other dimensions. Yet, we seem to function well in this universe with our limited perceptual representation.

The waking perceptual visual system can operate at non-conscious levels. The baseball player reacts to a hit ball with extraordinary non-conscious speed; bracketing possible trajectories, organizing complex muscle groups, addressing alternative possibilities and responses, and planning for next action in a seamless process that can have esthetic beauty. These processes are occurring on a non-conscious level that does not involve thoughts or feelings. It is not surprising that we are prone to confuse the concept of mind with the brain process of visual and perceptual integration, something increasingly within the capability of computer systems.

THE VISUAL REPRESENTATION OF SPACE

The major requirement for any perceptual system is consistency. The representation that we use in our mental process has to maintain an internal and external consistency for us to maintain a rational interaction to any reality outside ourselves. The fact that this representation consistently reflects external reality does not necessarily mean that the pictures in our mind are unique and realistic rendering of our external environment. In the Renaissance, artists and scientists discovered that much of our way of perceiving physical space is based on perspective. Artists developed a way of presenting spatial elements in a picture that was based on a consistent system. Perspective became "visual truth," the way in which three-dimensional space is represented on a two-dimensional canvas. A realistic appearance in depth and space was achieved by mimicking the exact geometrical relationship of how light would meet the eye in a three-dimensional scene (Fig. 12.1) (Mausfield, 2003). This led to a system of artistic construction rules for portraying space and depth in art that were later adapted through visual optics to explain image formation in the eye. Today, neuro-scientific studies of vision are often based on the affects that changes in viewing angle, height, motion, and distance have on the way that we see pictures. The rational for such studies depends on the assumption that perspective pictures in which the "eye is the camera" are the truest renderings of visual perception of space.

Linear perspective provides a consistency for the translation of three-dimensional space into two-dimensional representations in pictures (Hopkins,

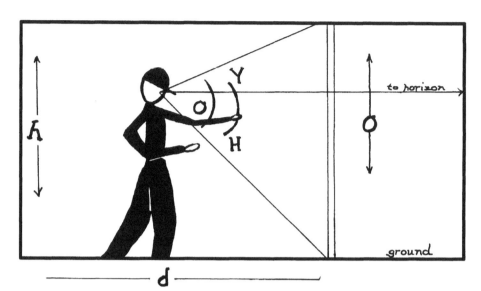

The horizon relations are given in terms of visual angle. The horizon ratio is
(tan Y + H)/tan H. The horizon distance relation is d=h ctn H.

FIGURE 12.1 Perspective – the eye as camera.

2003). Perspective as a component of visual representation is present in both
dream and imagery.

However, modern artists have developed a multiplicity of alternative schemes
for depicting pictorial space. Today, the idea that there is only one way to represent space in artistic renderings and photographs is considered old-fashioned and
potentially a matter of cultural bias or chauvinism (Hetch et al., 2003). It has also
been suggested that linear perspective is culturally based, a projection of Western
patterns for organizing the visual world (Collier & Collier, 1912/1986). Western
patterns of visual perception may be nothing more than highly selected stereo-
types that differ from those adopted by other cultures:

> "There is in our culture a common belief that vision has little or no context, that what we see
> is the result of a direct stimulus–response linkage between the image as stimulus and the cere-
> bral interpretation of stimulus. That is, that a direct connection exists between the external
> world and what we see, without intervention on the part of the culturally conditioned central
> nervous system. Ergo, vision is independent of experience – unaltered by experience [how-
> ever] vision like language, is not only structured but deeply contextual. As a consequence,
> once the grammar of vision has been mastered, it is possible to manipulate the 'meaning' of an
> image by manipulating the visual context of which the image is a part" (Hall, 1986).

These studies from anthropology call into question the larger neuro-scientific literature suggesting that the consistency of representation provided by rules of perspective are neurologically based. Neuroscientists tend to ignore the potential of
lived experience and cultural context to affect the consistency of our world-view,
and even to alter patterns of functioning in the human CNS.

IMAGERY

The CNS systems that are utilized in assessing our visual perceptions of the outside world are also utilized in imagery (Kosslyn, 1994). Imagery can be considered our ability to develop "mental images" of objects without actual perceptual input. Our conceptual interaction with the outside world is based on images and visual constructs. Visual imagery is based on the processing of visual signals carried out at CNS sites far from the primary visual processing areas involved in perception. Mental imagery differs from perceptual-based imagery in several basic ways:

1. It is slower and more controlled without the stimulus-based attention shifting that typifies external perceptual input.
2. Imaged objects fade quickly while external objects persist as long as they are present.
3. Images are limited by the information encoded into memory.
4. We have control over objects in images when the external world is rarely so cooperative (Kosslyn, 1994).

The cognitive architecture involved in image generation requires the activation of interacting memory processes. In imagery the visual representation of an object is achieved with either a global shape or part of a shape that our intrinsic memory utilizes as the representation of a particular object. Studies in infants suggest that our ability to interpret two-dimensional visual images reflects generalized physical principles that are hardwired rather than learned. Infants exploit the concept of cohesion in their recognition of objects – an object is considered a unified object only when composed of a series of surfaces in contact. Surfaces are considered to lie on a single object only when those surfaces are connected. This principle of continuity of objects means that we perceive objects as moving on connected paths from one place in time to another in non-intersecting paths because we perceive of such cohesive bodies as solid. Multipart images utilize spatial relationships between these representations of forms to create an image. The visual continuity of complex objects is maintained neurologically by at least seven different types of binding:

1. *Property binding*: Different properties such as shape, color, and motion are bound to the objects that they characterize.
2. *Part binding*: The parts of an object are segregated from the background and bound together.
3. *Range binding*: Particular values of a property such as color are defined within the dimension of that property (i.e., the visual spectrum of the color purple).
4. *Hierarchical binding*: The features of shape defining boundaries are bound to the surface defining properties of that object.
5. *Conditional binding*: The interpretation of one property (e.g., motion) depends on another (e.g., depth, occlusion, or transparency).
6. *Temporal binding*: Successive states of the same object are integrated across temporal intervals as in real or apparent motion.
7. *Location binding*: In which objects are bound to their locations (Treisman, 2000).

Visual object binding requires a cognitive interplay between perception, memory, and response. These spatial relationships based on object and category relationships often utilize a three-dimensional coordinate system in the representation of spatially related forms. It is possible that this coordinate system may be biologically defined (Damasio, 1994). It has been suggested that the consistent frequency of physiologic brain rhythms such as the hippocampal theta that occurs during rapid eye movement (REM) sleep could be utilized to define distance in cognitive space (O'Keefe, 1993). This is another among the several unproven but potential functions for the CNS electrophysiologic rhythms (Chapter 6).

By focusing our attention in imagery we can augment images, mentally drawing a location-based image through focused associations. For example, visualize the trunk of your car packed for a trip. We can shift our attention through such an image to look under and behind the bag of food on top, inside the suitcase to examine what we have packed. We inspect the objects in our images using perceptual-based techniques of encoding and interpreting. Pattern codes for objects, example a suitcase, are obtained through intrinsic memory.

Most imagery involves motion and requires the activation of CNS visual perception-based systems that recognize stereotypic patterns of object motion. The entire field of the image or a portion of the image can be altered. Our perceptual control in imagery includes the activation or completion of motions as well as the scanning and zooming of the image.

Motion can involve changes in both color and form. The brain registers the change in color first, before registering the change in direction of motion. This difference is not one of perception, but one of different processing speed required for each of these cognitive systems to integrate perceptual input (Zeki, 1999).

We each have a visual buffer that can "trigger" our attention visually, in image or in dream. This buffer may induce us to focus our attention on part of an image or lead us to disengage from a representation. Such triggers can be emotionally competent stimuli that can bypass the normal visual processing system allowing us to divert attention to those signals even when we are not paying close attention (Vuilleumier & Swartz, 2001). Perceptual and emotional input can be aspects of body states included in such neural brain "body maps" that can be triggered by associated visual imagery that change the focus and content of an image. Emotionally competent stimuli in such imagery can trigger associative memories that change our focus and attention to a specific part of the representation or to another representation that was not part of the initial image. Perceptual and emotional input can be aspects of body states included in such neural brain "body maps."

The process of imagery can be described as an operative series that includes the following:

1. *Picture*: Develop your pattern and configuration map of the image.
2. *Find*: Use attention to shift image properties and coordinate patterns.
3. *Put*: Focus on the description and relationship of a part to the whole image.
4. *Image*: Establish object names, size, location, orientation, and level of detail.
5. *Resolution and re-generation*: Delineate the comparative detail of this image.
6. *Look-for*: Integrate relevant memories.
7. *Scan, zoom, pan, and rotate*: Your presence, your operative attention in the image.
8. *Answer-if*: Do properties associated with the image answer an already developed cognitive search parameter? (Kosslyn, 1994)

This imagery operative cascade is at least a partial descriptive paradigm of the cognitive processes involved in the cognitive process of imagery – visual processing occurring without actual perceptual input.

Dream images, although they may appear to be images from the real world are our neural translations presented as pictures in coded three-dimensional space. Dream imagery is as actually as thin as "thought" and the thought space of such imagery is very different from perceptions of real space (States, 1997). Perceptions based on images from the external world have a vitality and vivacity that imagery often lacks (Scary, 1995). As Jean-Paul Sartre (1991) observed, if the object we select to imagine is the face of a close friend, one known in intricate detail, it will be, by comparison with an actual present face, "thin," "dry," "two-dimensional," and "inert." Analyzing the cognitive process involved in visual imagery parodies the visual experimentation of the impressionist and post-impressionist artists. It was Picasso that stated, "it would be interesting to preserve photographically ... the metamorphosis of a picture." Possibly one might then discover the path followed by the brain in materializing a dream (Leger, 1938).

It is difficult even to define what there is of a neural representation of the external environment that can be represented visually as a picture: "A sign is a picture if the perception of the essential properties that the sign has in relevant respects is identical to the perception one would have of the corresponding properties of some other object under a certain perspective and if this perception is constitutive for the interpretation of the sign" (Sachs-Hombach, 2003). The reduction of imaginative "seeing" to a brain-based operative cascade is but one way of approaching image creation. This operative cascade describing imagery formation can be artificially created as well as adapted to the process of dream imagery formation. It is an approach that can be adapted to both artificial intelligence (AI) systems and cinematographic approaches to producing images that reflect story and plot (Fig. 12.2).

DREAM IMAGERY

Most dreams are remembered as images. Dreaming like imagery does not generally incorporate perceptual input. Sleep by definition is a reversible perceptual disconnection from our environment, a state of perceptual isolation. The dreams of sleep are rarely altered by perceptions of the external environment, particularly visual stimuli. Yet most of us dream in images. Dream imagery differs from the imagery of waking consciousness in the ability to quickly "grab" our mental attention, reminiscent of the way it is grabbed by stimuli during perception (Kosslyn, 1994). Dream images also differ from waking images in that they are generally out of our conscious control. In this way dream imagery resembles the projected imagery of film in which the spectator "gives over" conscious control of perceptual input to the filmmaker. There are other similarities between dream and film imagery. The specific images presented on film are often hard to remember in the same way that dreams are hard to remember. Movies can be hard to remember, as hard to remember as the actual events of yesterday, or last nights dream. You find yourself remembering moments of a film, trying to find your way back to the mood, and associations associated with that particular moment of cinema. Film like dream can tap into emotionally competent stimuli that lead to personal associative memories that take the viewer down story paths much different from those

FIGURE 12.2 Operative task – create an artificial dream.

planned by the filmmaker. Like dreams, certain moments of films viewed decades ago will nag as vividly as moments of childhood (Cavell, 1971).

Imagery like dreaming, is involved in the creation of visual images that are not based on real-time perception, and like dreaming is primarily visually based and replete with multiple associations (Marks, 1990). Much of dreaming imagery is viewed from the perspective of the dreamer, occurring to the left, to the right, in front of, or behind the dreamer's point of perspective in the dream. Self-awareness in dreaming is often called "lucid" dreaming in that the dreamer present in the dream has at least some self-awareness and control of the dream process (LaBerge, 1985). Since most dreams are egocentric in nature and occur from the perceptual point of view of the dreamer, the dreamer is present in almost all dreams. This egocentricity is generally expressed as a general sense of one's body as a bounded object located within a space containing other objects. Such egocentric awareness in dream requires some level of "lucid" self-awareness. From this perspective most dreams can be considered lucid in that the dreamer is present in the dream and in control of the dream imagery.

This imagery operative cascade is also part of dream imagery with dreaming incorporating the same perceptual operating systems that are utilized in imagery. Even a form of sub-conscious visual buffer is present in dream. This visual buffer can include emotional and experientially based memories capable of triggering our attention. Let's return to the image of a packed trunk. If you use to climb and your brother has died in climbing accident, the image of a rope packed your dream trunk would grab your attention and trigger your visual buffer focusing leading you to an emotionally competent set pattern of memory. Extending this painful allegory from the simple image of a rope in the trunk of your car into negative emotions, judgment, and quilt – the trunk of your car may not have been a good place for your brother's rope when he was out on the glacier climbing across crevasses. Such dreams have power and significance. They are very likely to be remembered the next morning and affect your feelings and mood during that day.

ARTIFICIAL DREAM IMAGERY – FILMMAKING

Dreams incorporate the perceptual operating systems of imagery. Because of this, there are perceptual characteristics of dream imagery that can be computer generated. Images created on film are likely the closest cognitive parallel to dream imagery. Both are visual images that are not under the conscious control of the viewer. Both are coded representations of an external reality. Filmmakers have learned how to fool our perceptual visual system into perceiving a continuous moving picture with three-dimensional characteristics of depth and foreground using a projected series of still two-dimensional images. Filmmakers understood what works and does not work in the process of creating moving pictures long before scientists had a clear understanding of how the visual system operated. Scientists were to eventually find that film as an artistic medium utilizes many of the same processes used in the cognitive organization of visual experience. We will follow this correlate between film and dream imagery more closely in a later chapter.

VISUAL COMPONENTS OF DREAMING

Vision like emotion is a brain-based component of dreaming clearly amenable to brain-based scientific study. Neuroscientists understand the neuroanatomy and the operative processes involved in the visual perceptual system better than any other perceptual system. Vision may be the best described of any cognitive processing system. Variants of the visual system are present even in primitive invertebrates. Neuroscientists can use a wide variety of animal models in their studies, some using visual processing systems that are far different from the human norm including the multifaceted optics of insects, altered wavelength and color sensitivity as well as extreme focus capabilities in birds, and extended motion and light capabilities in many species. Techniques such as PET scanning, neuropathologic studies based on disease and trauma, animal lesion, and single neuron studies have been used to help us to understand that a large portion of the CNS is involved in visual perceptual processing. The neural process of visual perception has turned out to be one of the easiest brain-based cognitive systems to study. This system has obvious evolutionary and survival value. We use vision to function and maintain our waking existence in the external world. Artificial constructs of these visual perceptual systems have been developed and utilized in popular culture in photography, film, television, and in computer simulations. Neuroanatomically, we understand these systems fairly well – better than we understand the majority of other cognitive processes. The evidence indicates that a majority of cognitive visual processing systems are brain rather than mind based.

Film is an artificial construct that approximates the same visual imagery system utilized cognitively in dream. The visual imagery system interacts with the emotional processing systems that are also present in dreams. Film images can affect the viewer much as dream imagery affects the dreamer. Images that are often out of the control of the viewer can trigger emotionally competent stimuli that induce both the dreamer and the viewer of film to connect with associated and sometimes hidden memories. Emotionally powerful images are remembered, altering the way in which both the waking dreamer and film viewer relate to the external world.

The images of dreaming, those titanic waves and dragons of dream have clear brain-based correlates. Dream imagery may not truly be an aspect of mind. Images from the imagination are built in operative constructs. Images from waking imagery and dream have a two-dimensional thinness, a lack of complexity, and depth that suffer when compared with our waking perceptual experience. We should not be surprised to find that our visual perception of the waking sensory world surpasses our visual capabilities in the imaginary realm.

CHAPTER 13

Thinking and Dreaming

We like to think of our minds as containing chains of thought, or streams of consciousness, as though they were orderly arrangements of linear events, one notion leading in cause-and-effect to the next notion ... I can acknowledge openly that my own mind is, at most times, a muddled jumble of notions, most of them in the form of questions, never lined up in any proper order to be selected and dealt with when the time allows, most of the time popping into my head unpredictably and jostling against any other ideas that happen to be floating along, each new disturbance amplifying the disorder of all the others, creating new geometric shapes of chaos imposed on chaos

(Thomas, *The Fragile Species*, 1996)

What the human central nervous system (CNS) does best is think. We think when we are awake. We think when we dream. We think when we sleep. When we are conscious, we think. Whether awake or attempting sleep one of the most difficult things for us to attempt is to turn off the process of thinking. But what is thinking? In order to differentiate thinking from processes that appear to be thinking, but may be something else, once again we are forced to examine what the definition may be for a cognitive state difficult to define. The dictionary definition derives from a root meaning of "causing something to appear to oneself." Currently, accepted definitions for thinking include "to exercise the mind upon, to consider, attend to mentally or apply the mind to," "to form or have an idea of a thing in one's mind," and "to think of something in the way of a plan or purpose" (The Oxford English Dictionary, 1971). Mark Belth (1977) in his book *The Process of Thinking* defines thinking as the "act of following out, and examining at the same time, a path, pattern, mapping, form, or formula until what has been called for in that map, path, pattern, form, or formula has been concluded and the whole of it has been considered for its inner and outer consistencies and its warrantable circumstances." Stated more succinctly, thinking is an act that includes reflection upon itself (Belth, 1977). Martin Heidegger, among modern philosophers, has focused his work on the process of thought. He notes in his series of 1944 lectures published as *What Is Called Thinking* (1968) that the human species is naturally inclined to think. He emphasizes the nature of genuine thinking: "Multiplicity of meanings is the element in which thought must move in order to be strict thought" (1968). He points out that scientific knowledge is based on a methodology that excludes alternative explanations. As per Heidegger's definition, science itself may not arise from thought. Benson (1994) in his book *The Neurology of Thinking* reviews the plethora of definitions before deciding to define thinking as *representing the activities of a number of diverse, precisely interrelated nervous system functions that process thought contents*. He rigidly separates *thought content* (all data received and accumulated by the individual) from *thought processing* (the acts of receiving, perceiving, comprehending, storing, manipulating, monitoring, controlling, and responding to that data). Benson has adopted this approach in an attempt to more clearly differentiate the brain-based correlates of perception from the processes of "thinking" about that perceptual input.

THINKING IN DREAMS

Dream thought differs from waking thought:

> Focused waking thought involves serial processing, one thing after another in an orderly sequence. It cannot adequately explore or "appreciate" the distributed massively parallel nature of our memory nets. But dreaming ... is able to do this due to [a] linked group of properties: hyperconnectivity, including condensation, emotion-guided contextualization, pulling together related contexts; pulling things together into a picture or pictured metaphor. Dreaming is less serial, more parallel. It deals with "lots of things at once" rather than "one thing after another" (Hartmann, 1998).

Dream thoughts are considered by some to have a fragmentary storyline that is periodically disrupted by bizarre intrusions (Hunt, 1991). This postulate has led to the proposal that thinking in the dream state is logically impaired (Mamelak & Hobson, 1989). It is suggested that thinking during dreaming is sleep state dependent through the night with hallucinatory, bizarre thought occurring in association with rapid eye movement sleep (REMS) (Fosse et al., 2001). This hypothesis is based on the definition of thinking as focused thoughts involving internal deliberation, and hallucinations defined as endogenous percepts disconnected from perceptual reality (Fosse et al., 2004). Although these studies are interesting, study results are presented based on the authors' theoretical bias toward the REM sleep being the primitive mind's basis for dreaming. Their actual findings sometimes contradict their theories even when based on personally developed scales for assessment of nocturnal thinking. Participants in these studies were asked: "Would your thinking when awake be the same as it was in the dream if the event that occurred in the dream occurred while awake?" and "Would your thinking when awake be the same as your thinking in the dream regarding the occurrence of the event itself?" (Kahn & Hobson, 2006). These individuals report that thought in their dreams is somehow different from their thoughts when awake. This data is interpreted to demonstrate that "Thinking about the scenario itself (dreaming) is very different from wake state thinking." Later the authors seemingly contradict their conclusions stating, "Cognition within a dream scenario is similar to that of wake state cognition" (Kahn & Hobson, 2005). These studies, are interpreted as support for activation–synthesis and the association of REMS with bizarre, hallucinatory like dream mentation. The possibility is ignored that other variables and methodology are liable to affect results. Sleep stage based variability in dream thought and content remains a debatable proposition.

Waking thought is generally considered to be more precise than dream thought, with that preciseness based on the constraints and interactive rules of waking life. It is suggested that waking thought may be more "rational" than dreaming thought, with rational thought being the individual's ability to notice and differentiate aspects of personally subjective experience from known objective data (Shoben, 1961). Waking thought is, however, varyingly rational. Stanovich and West (2000) note that hundreds of studies on waking thought demonstrate that:

> people's responses often deviate from the performances deemed normative on many reasoning tasks. For example, people assess probability incorrectly, they display confirmation bias, they test hypotheses inefficiently, violate the axioms of utility theory, they do not properly calibrate degrees of belief, they project their own opinions on others, they allow prior knowledge to become implicated in deductive reasoning, and they display numerous other information processing biases.

Wolman and Kozmova (2006) have developed an operational definition for rational thought as, "a mental process that utilizes an individual's internal logic and is based on an idiosyncratic belief system. The mental process of rational thought intervenes between sensory processing and the creation of meaning, and leads to a conclusion or to taking action." These authors identify a series of categories of thought process associated with rational thought:

1. *Analytical*: Comparing and contrasting, evaluating, reason, logic, reflection, and contemplation (27.2%).
2. *Perceptual*: Paying attention to visual, auditory, gustatory, olfactory, tactile, and kinesthetic occurrences (22.4%).
3. *Memory and time awareness*: Remembering and recall. Recognition of characters, history, events, abilities, time, and the dreaming state (15.0%).
4. *Affective*: Distinguishing, naming, and/or verbalizing the spectrum of feeling and psychophysiological states (9.5%).
5. *Executive*: Decision-making, problem solving, planning, and agency (7.3%).
6. *Subjective*: Personal history, characteristics, appearance, beliefs and desires, skills and goals (7.3%).
7. *Intuitive/projective*: Assumption, lack of sufficient facts, and erroneous attributions (6.4%).
8. *Operational*: Reading, writing, counting, and measuring (4.7%).

Wolman and Kozmova analyzed a series of dream reports and found these processes of rational thought present in many dream reports (percentages of dreams including types of rational thought are in brackets). This suggests that even when dreams appear bizarre or hallucinatory to the dreamer, the thought processes that support the percept of the dream can appear rational to the outside observer. This may be because dream reports utilize the standard rational structures that each of us use in interpreting and presenting to others our personal experiences and interactions.

BIZARRENESS IN DREAMS

Are dreams really bizarre? They tend to be rather plodding by fictional standards (States, 1997). Bizarreness as a mode of thinking differs from the "natural Newtonian laws of wakefulness" that characterize waking reality and the conventional continuity of waking life (Bonato et al., 1991). Dreams that fail this reality test are often characterized as bizarre. Bizarreness in thought can be measured using content scales that plot incongruities, uncertainties, and discontinuities. Incongruities are mismatches – orange skies at midday. Uncertainty refers to the non-specificity for persons, things, and events in the dream such as unidentifiable localities or characters. Discontinuities are abnormal shifts in person, place, or action – the dreamer becomes someone else or moves abruptly to a different locale. Studies suggest that such "bizarre" events may be more common in dreams than in focused waking thought (Flanagan, 2000). Because of these "bizarre" characteristics, dreams are considered to be less efficient and reliable than waking thought. This may be true were the dream the basis for waking activities, however, in its own dream world, are these characteristics of dreaming actually inefficient and unreliable? In the dream world there is no need to prove that a dragon actually exists. There is no need to prove that the inner world of dreams has a rational, provable continuity with the external world. The standard criteria that we use to test capability in thought or knowledge in our waking world apply poorly when used to analyze the dream state. Because of the personal focus of dream, it is also difficult to rely on rating scales or outside judges to assess such a difficult content criteria as "bizarreness" (States, 1993a).

The degree of bizarreness in dreams has been shown to correlate with performance on various measures of creativity and imagination (Hunt, 1989). Dream thought is much like the imaginative thought and the inner speech required for composing art, doing science or mathematical problems, or thinking about the possibility of future alternatives. These states have in common with dreaming the phenomena of bizarreness in image formation and the radical discontinuity that is often required for creative process. These states require a cooperative competition between linear and non-linear thought much as is to be found in dreaming (States, 2000). Most artists are not dreaming or delusional and know that there is a difference and therefore a dualism between art and reality (Arieti, 1976).

ASSOCIATIVE THOUGHT IN DREAMS

In the dream, we may be at the site of pure creativity where the associative process involved in original thought is open to our examination unhampered by matters of comprehensibility or conceptuality. The associative process of dream has historically been a focus in psychoanalysis where what matters is less the dream than the dreamer's waking associations with dream content. Dream thought differs from waking in that dream thought is associative, albeit based on rather unusual rules of association. The images of dream are always dynamically changing requiring nothing more than the impetus of association to lead from one image to the next. In the dream state, potential multiple associations occur independently of the serial and time-based nature of experience (States, 1993a). The dream is not a series of images attached by causality, but a series of images demonstrating an evolving process of association (States, 1993a). Dream images follow each other propagated in an associative flow based on connection and syntactical appropriateness (Overstreet, 1980). This process can become increasingly and beautifully complex. As Sartre pointed out "a dream … is a world" (1968). Dreams have no choice but to represent what is occurring in the dreaming brain. This is an organized process. "A dream object does not transform randomly into another object, but into an object that shares formal associative qualities with the first" (Rittenhouse et al., 1994). Dream coherence develops based on the fact that the associations recalled are those which have significant importance for the dreamer such as emotionally competent stimuli with emotional affect (States, 2000). A function of dreaming may be to provide this kind of associative thinking that we have difficulty in doing while awake.

THE NEUROLOGICAL BASIS OF THOUGHT

A primary function of the CNS is to think, something that most of us do constantly when awake and in much of our sleep. There is little neuropathological or scanning data supporting the hypothesis that discrete sites of the brain are involved in the process of thought. Specific actions based on thought will light up large portions of the CNS on positron emission tomography (PET), single photon emission computer tomography (SPECT), and functional magnetic resonance imaging (f-MRI) scans. If you think of moving your left little finger, a portion of the right motor cortex shows intense activity. If you visualize an image from your childhood, portions of visual cortex and hippocampal memory sites will be among the sites showing activity on scans. Thinking, however, appears to be a global CNS process without clear areas of CNS localization except when associated with specific perceptual or motor activities. Benson (1994) after conducting an intensive search for neuroanatomic correlates to explain the process of thinking eventually states that, "Higher mental controls are supramodal in operation."

The conceptual classification of "disorders of thought" includes such diagnoses as schizophrenia and suggests that malfunction of some portion of neural processing should be responsible for the diagnosis (Benson, 1994). While some schizophrenic patients may show general non-specific areas of decreased CNS frontal activity, this finding is neither specific nor diagnostic for the many forms of this psychiatric disorder. As Benson points out "… no single anatomic site or sites underlie thought disorder" (1994).

Some practitioners of meditation can accomplish states of "non-thought" or at least "non-focused thought" through their meditative practice. Some practitioners studied with f-MRI scans demonstrate lower levels of frontal cortex brain activity during meditative states (Carter, 1998). The best evidence for CNS changes produced in these practitioners is electrophysiologic based on the increased frontal gamma activity noted in long-term practitioners of Tibetan Buddhism (Lutz,

2006). This general and global CNS change occurs with long-term training in the development of "non-focused" thought capabilities.

THINKING AND THE MIND

Researchers have had difficulty finding the neuroanatomic sites in which thinking occurs. Some neurologists have come to view thinking as a supramodal process occurring in the higher cortex without an association with specific neuroanatomical sites (Benson, 1994). The postulated interconnecting cognitive processes with their associated cortical anatomical sites can be diagrammed conceptually (Fig. 13.1).

Figure 13.1 describes the hypothetical integration of perceptual input that defines extrapersonal space into the internal milieu of the CNS. Within a matrix of integrated neural networks, multiple activities are proposed to be occurring simultaneously linking together simultaneously operating subunits (Benson, 1994). This may be the complex cortical processing of thought required for in the integration of perception into consciousness. Data supporting the existence of such a system is limited.

Neurocognitive pathologic data from individuals with extensive frontal brain damage indicates that prefrontal cortex damage does not interfere with the ability to perform cognitive tasks (Benson, 1994). Such negative findings may be due to the fact that the behavioral abnormalities observed in brain damaged patients do

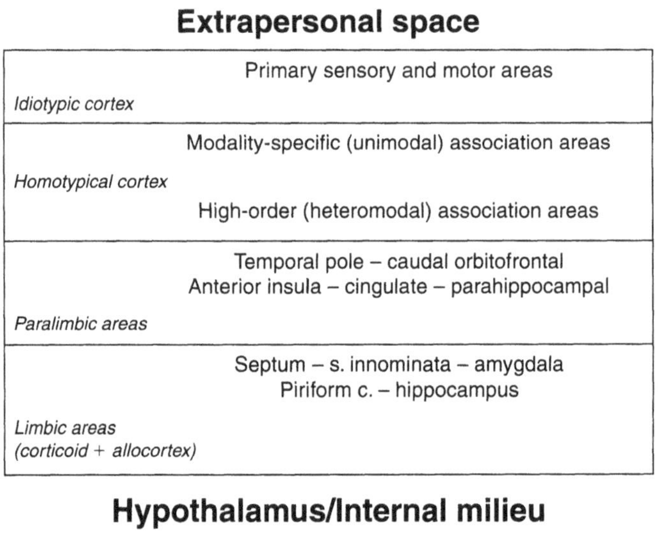

Extrapersonal space

Idiotypic cortex	Primary sensory and motor areas
Homotypical cortex	Modality-specific (unimodal) association areas
	High-order (heteromodal) association areas
Paralimbic areas	Temporal pole – caudal orbitofrontal Anterior insula – cingulate – parahippocampal
Limbic areas (corticoid + allocortex)	Septum – s. innominata – amygdala Piriform c. – hippocampus

Hypothalamus/Internal milieu

FIGURE 13.1 Mesulam's diagrammatic representation of the varieties of human cerebral cortex with differentiation into idiotypic (primary), unimodal (secondary), and heteromodal (tertiary) association cortices and paralimbic and limbic cortices (1985).

not actually reflect damage to the finite area of brain involved; rather, they reflect the ability of the individual's nervous system to compensate for the damage (Jackson, 1932). It is also possible that our current scientific techniques used in neuropathological analysis are not actually measuring or analyzing the underlying abnormalities involved in thinking or its associated disorders.

Our original definition for mind in this book was that mind included those cognitive processes that we cannot measure or define with our technology. Thought is one of those processes. There is little data to support the association of specific neuroanatomical sites or pathways associated with this cognitive function. Thinking has no clear brain-based correlates. In studying thought, we are considering an area of CNS functioning that is poorly explained by our current technological capabilities. Of the cognitive processes that we have considered, thinking appears to be the cognitive process that is most clearly mind.

Visual imagery and emotion clearly have brain-based components. In studying dreaming, we have a brain-based understanding of the cognitive processes of visual imagery and emotion that are major components of the dream. Only at the limits of scientific understanding do we approach areas of these cognitive processes that we cannot explain as brain-based components of those processes that may be mind. The mind-based component of dream turns out not to be the incredible visual imagery. Mind is not to be found in intense impactful emotions. Mind in dream appears most clearly in the contents of dream thought.

The Cognitive Organization of Dreaming

WHAT WE KNOW

We understand much of the biochemical, neuroanatomical, and electrophysiological basis for rapid eye movement (REM) sleep. If we restrict our definition of dreaming to REM sleep, we understand the neural framework of dreaming. We understand much of the biological basis for vision, and emotional processing, and something of the brain-based process of memory. Emotion, memory, and visual imagery are primary characteristics of dream. If the cognitive process of dreaming includes only memories, emotions, and visual imagery, we understand much of the biological basis of the cognitive process of dreaming.

It has been proposed that mind includes any cognitive process that occurs without waking perceptual input (Damasio, 1994). Imagery is then an aspect of mind, based on its lack of actual perceptual input. Reversible perceptual isolation is the definition of sleep. From this perspective, whatever occurs in sleeping and dreaming is conceptually mind. If this is so, we can be said to understand mind quite well. Our definition of mind becomes the cognitive integration of visual perception, memories, and emotion occurring during states of perceptual isolation. Based on this perspective, the neural interactions required for respiration to occur during sleep is mind (Hobson, 1994).

This is a reductive and simple view of the mind. Couple this definition of mind with the realization that on many levels, perceptual processing can parody the processes of conscious cognition (Chapter 11) and you have achieved a conceptual view of consciousness that parodies the automaton, the artificial intelligence (AI) reactive to the environment with motive power, but without active intelligence. From this perspective mind equals brain, despite contrary evidence. Cognitive function is based on brain function. Thinking, thought, feelings, esthetics, spirituality, and creativity; the processes for which we currently have difficulty finding brain correlates, await only the arrival of better analytic technology.

Unfortunately, this perspective of brain equaling mind in the process of conscious automation may leave out whatever there is in our conscious functioning that makes us human. What is there about us that is more than a physical automaton? Perhaps we are conscious robots, and the prospect of functioning AI is just around the corner. Perhaps dreams are nothing more than internalized perceptual hallucinations. Perhaps dreams are meaningless.

If we understand how memory, emotional processing, and visual imagery of dream work, we understand the biological basis of some of the major cognitive processes involved in dreaming. We understand the parts of dreaming that are not mind. We can use the dream to examine what remains, looking through this window provided by the scientific study of the dream to examine the borders of the mind.

THE FORMAL STRUCTURE OF DREAMING

The cognitive process of dreaming has a formal structure. This formal structure is required to organize dream thought and images in order that dream content somehow makes sense in waking consciousness. The similarity of dreaming to other experiences of non-perceptual visual imagery has led some neuroscientists to assume the perspective that dreams are a form of visual hallucination that occurs during sleep. This perspective has both limits and advantages.

The formal structure of the waking integration of dream is through narrative and story, the basic processes through which we organize experience. The narrative structure of dreaming reflects our attempts at comprehension of our world. One way of viewing the dream is as a demonstration in an unadulterated form how we organize perceptions and memories in order to make sense of the world.

The formal structure of most dreaming is visual. Cinematographic techniques imitate the way that we visually organize our perceptual experience. Film attempts to parody the perceptual/neurological movies that we create in our own minds utilizing the same perceptual system utilized in waking and dream imagery to affect us emotionally and alter our perceptual view of the exterior universe.

In this segment we will examine these projections of dream. How dreams are both similar and different from waking hallucinations. How film can be the projected dream. Through film we have the ability to create artificial dreams and emotionally manipulate the viewer by parodying the emotional processing system active during dreaming. How dreams are stories. These turn out to be large questions. Questions about what it is to be conscious. Questions about what it is to be human.

CHAPTER 14

Hallucinations and Dreams

Is this a dagger which I see before me,
The handle toward my hand?
Come, let me clutch
      Thee:
I have thee not, and yet I see thee still.
Art thou not fatal vision, sensible
To feeling as to sight? Or art thou but
A dagger of the mind, a false creation,
Proceeding from the heat oppressed brain?

(Shakespeare, Macbeth)

In order to function in society, each of us is required to have the ability to differentiate between dreams and reality. Otherwise it is possible that we might be removed from society, or at the least, heavily sedated. Delusions and hallucinations are included in the diagnostic criteria for a variety of psychiatric illnesses, primarily the psychoses (DSM-IV, 1994). A hallucinatory experience was originally defined as a perception without an object (Esquirol, 1994). The cognitive process of visual imagery also involves seeing of something that is not externally present, however, most of us do not think that our visual imagery is a direct reflection of external reality. The individual having a hallucinatory experience accepts that the hallucinatory experience is referring to an external reality. The hallucination is believed to be an experience of normal perception (Reed, 1972). Since Esquirol's original definition could be construed to include both dreaming and imagery as hallucinations, Jaspers (1923) developed a definition of hallucination as a discrete category of experience independent of imagery: "Hallucinations proper are actual false perceptions which are not in any way distortions of real perceptions but spring up on their own as something quite new and occur simultaneously with and alongside real perceptions." A hallucination has perceptual characteristics that occur in the absence of external experience. The subject experiencing a hallucination feels that it has reference to external reality although objectively it does not.

Reed (1972) has attempted to clarify in psychological terms the differences between imagery, illusion, and hallucination. Imagery is a cognitive process based on the perceptual reconstruction of stored material. Illusion involves the misrepresentation of input in terms of its synthesis with stored material. Hallucination involves the perceptual reconstruction of stored material and its misrepresentation in terms of input. He points out that the only critical characteristic of what we call a hallucination is the conviction the hallucinator has during the experience that the experience is reality. Subsequently to the experience of a hallucination, the subject is loath to abandon this conviction. "In other words, hallucination is crucially a question of reality testing" (Reed, 1972). For the psychiatrist attempting to use this strange symptom in diagnosis, hallucinations are considered to be a sign of psychosis only when the individual believes the perceptual hallucination to be real (DSM-IV, 1994).

Hallucinations occur in different perceptual modalities with the different perceptual types of hallucinations associated with different medical and psychiatric diagnoses. Visual hallucinations are typically associated with chronic schizophrenia and acute organic states that induce a clouding of consciousness. Organic disorders that cause hallucinations include substance-abuse disorders, medication psychotoxicity, ocular and olfactory pathology, neurological, endocrine, and metabolic disorders (Asaad & Shapiro, 1986). Classic examples are the "pink elephants" described during the delirium tremens of ethanol withdrawal. Organic hallucinations are more likely to be repetitive, fragmentary, and stereotypic, and tend to be extensions of misperceptions (shadows perceived as ghosts). Correction and reassurance can sometimes eliminate hallucinations in organic patients. The visual hallucinations of psychiatric illness rarely respond to correction or reassurance. The visual hallucinations sometimes present in chronic schizophrenics are often fixed and developed into elaborate systems (Perry & Markowitz, 1988).

The visual hallucinations induced by psychedelic drugs differ from the hallucinations of organic and psychiatric illness. Psychedelic hallucinations tend to develop as flickers, flashes, spot of light, and streaks of color. These visual changes may develop into geometric forms and repetitive patterns, in some individuals becoming increasingly complex with isolated pictorial elements. True visual hallucinations do not show such progressive development, typically appearing suddenly and fully formed without progression to other levels of organization (Reed, 1972). True visual hallucinations are false images occurring with eyes open in a lighted environment. Such true hallucinations can be differentiated from pseudohallucinations, experiences appearing in inner subjective space that the subject knows not to correspond with external reality (Reed, 1972). Visual images that are reported as occurring when the eyes are closed, as well as hallucinations occurring as hypnagogic or hypnopompic phenomena at the initiation or discontinuation of sleep are not considered to be true hallucinations by many authors (Goldman & Foreman, 2000).

Auditory hallucinations experienced as noises, music, or most commonly voices, are typical of schizophrenia. Auditory hallucinations are reported by 28–72% of patients with the diagnosis of schizophrenia (Black et al., 1988). Olfactory hallucinations can be symptoms of schizophrenia, epilepsy, and depressive illness. Tactile hallucinations are characteristic of drug-induced acute organic states, particularly those induced by cocaine. Schizophrenics can have a wide variety of perceptual hallucinations including gustatory (taste) hallucinations as well as pain, proprioceptive (balance), and kinesthetic (motion) sensations.

Hallucinatory experiences are sometimes reported by normal individuals without medical or psychiatric disease. Visual imagery and auditory hallucinations are often reported at sleep onset (hypnagogic hallucinations) and when waking up (hypnopompic hallucinations). In one study of college students, 63.18% reported having experienced hypnagogic hallucinations, with 21.42% reporting hypnopompic events (McKellar, 1957). Hallucinations have been reported from sensory deprivation experiments. After long periods of diminished or monotonous and unpatterned sensory input, many subjects report flashes of lights, simple patterns or wallpaper-like phenomena with a few perceiving fully integrated scenes (Heron et al., 1953). It is unclear as to whether such induced hallucinations occur secondary to the sensory deprivation or are hypnagogic phenomena occurring because the individual is unable to remain awake in the deprived sensory environment. Hallucinations can also be induced through sleep deprivation, severe malnutrition, exhaustion, and the delirium associated with infectious illness, exposure, or toxic states.

In our culture as well as in shamanistic Native American, Asian, and South American tribal cultures, individuals may deliberately attempt to induce hallucinatory experiences. These spiritual and cultural rituals predate recorded history and may be based on rituals utilized by Paleolithic hunters (Eliade, 1964). The Native American tribes of North America utilize induced hallucinations in attempts to attain spiritual insight. Such rituals occur within a cultural context and are utilized after training in the spiritual traditions of the group. Most shamanistic traditions utilize the dream content reported by initiates in such rituals (Eliade, 1964).

Cultural traditions that include a belief in visions can predispose individuals to report nocturnal hallucinations from sleep. In ultra-orthodox Israeli Jews whose traditions include a belief in demons, particularly dead souls that visit during the night, more than 40% of young men undergoing psychiatric evaluation report such hallucinatory experiences (Greenberg & Brom, 2001).

DREAM-LIKE PHENOMENA

There are nocturnal experiences that can appear to be dreams based on a report of mental content, usually frightening in type, that is reported from sleep. There are rare cases of seizures presenting only as "nightmares." These are almost always reported from patients with either daytime seizures or with a history of central nervous system (CNS) disease.

Nocturnal panic attacks can occur either during or immediately after nocturnal awakenings from NREM sleep, usually in the first 4 hours of the sleep episode. The frequency of such panic attacks is correlated with patient report of nightmare frequency and many patients report that dysphoric dreams precede their attacks. Yet there may be no dream recall reported on awakening other than the immediate realization of occurrence of a panic attack. These panic attacks occur most commonly in patients with anxiety disorders and waking symptoms of panic. Dream-like events can be reported from patients with diagnosed dissociative disorders. The DSM-IV (1994) defined dissociative disorders include dissociative identity disorder (formerly called multiple personality disorder) and dissociative fugue. Individuals meeting waking criteria for these diagnoses may at times experience as a "dream" the recall of actual physical or emotional trauma.

Polysomnography studies indicate that the "dreams" are actually occurring during periods of electroencephalogram (EEG) documented nocturnal waking. These "dreams" are actually awake dissociative experiences and can be considered hallucinations in that the patient is awake, and experiencing perceptional imagery without actual perceptual input. The individual may believe these experiences to be real (Pagel & Nielsen, 2005).

PARASOMNIAS AND UNUSUAL DREAMS

Parasomnias are defined as undesirable physical events or experiences that occur during entry into sleep, within sleep, or during arousals from sleep. These experiences can include sleep-related movements, autonomic motor system functioning, behaviors, perceptions, emotions, and dreaming. Parasomnias become clinical diagnoses when they result in sleep disruption, nocturnal injuries, waking psychosocial effects, or adverse health effects.

The arousal disorders are parasomnias characterized by either partial or complete arousal from deep sleep. Sleepwalking, confusional arousals, and sleep terrors can be exacerbated by sleep deprivation, sleep fragmentation, and psychological factors including anxiety and stress. All share the clinical characteristics: occurrence in the first half of the night; confusion/autonomic behavior; difficulty waking from the event; fragmented imagery; rapid return to sleep; and amnesia of the event (Chapter 11). Sleepwalking (somnambulism) is most frequently characterized as quiet wandering around the home. Somnambulism is not usually associated with injury, and most spells are brief (5–15 minutes). During a confusional arousal, the child may seem to be awake, if not considerably confused. The child may cry, yell, moan, or speak in unintelligible sentences and may or may not recognize the parental figure. A "blood-curdling" scream and autonomic discharge accompanies the more extreme form of confusional arousal, the sleep terror.

REM (rapid eye movement) sleep behavior disorder (RBD) often presents with a history of violent, explosive movements, and nocturnal injuries. RBD is a REM sleep parasomnia most common in late middle-aged males and patients with chronic neurological conditions such as Parkinson's disease. RBD episodes occur in patients experiencing a failure of the motor paralysis that is a normal physiologic characteristic of REM sleep. RBD patients act out their dreams. The RBD experiences that are likely to result in injuries to the sleeper or bed-partner often occur in association with frightening dreams. The most commonly reported dream experience in RBD patients involves being threatened or attacked by unfamiliar people or animals.

Isolated sleep paralysis with hypnagogic hallucinations is a REM sleep parasomnia that is also secondary to a dysfunction in the motor paralysis associated with REM sleep. Disturbing hypnagogic hallucinations and frightening dream sequences accompany a paralytic state on awakening from dream in which the patient cannot move. These differ from nightmares. Nightmares, although they may involve some degree of movement inhibition (e.g., being unable to run from an assailant) or some degree of apparent wakefulness (e.g., becoming aware during the nightmare that it is, in fact, a nightmare), are usually not accompanied by feelings of total paralysis.

The dream-associated behaviors of RBD and sleep paralysis are more common in individuals with obstructive sleep apneas or periodic limb movements that result in increased awakenings during the night. These are sleep-related behaviors and experiences over which the sleeper has no conscious deliberate control. The parasomnias, especially REM behavior disorder, have been associated with nocturnal injuries to the sleeper and bed-partner, and have on occasion been the focus of legal proceedings after nocturnal injuries or deaths.

ARE DREAMS HALLUCINATIONS?

Dreams are not true hallucinations. They fail the reality testing criteria. It is the very rare individual that cannot distinguish dreaming from the experience of external reality. Dreams do have hallucinatory characteristics in that they are perceptual experiences without external object, appearing in an inner subjective space that the subject knows not to correspond with external reality.

Some dream theorists believe that dreams are a form of visual hallucination. This perspective is based on formal characteristics of the dreaming process that can be considered hallucinatory: visual and motor hallucination, the delusional acceptance of hallucinoid experience as real, extremely bizarre spatial and temporal distortion, strong emotion, and the failure to remember. These authors suggest that the dream is a hallucination because the dreamer has a "delusional" acceptance during dreaming of the dream experience as being real (Hobson, 1988). This approach avoids the reality testing criteria for hallucinations by divorcing the dream state from waking.

There are obvious problems with the assumptions on which this theory is based. As noted in the last chapter, it is debatable as to how bizarre dreams actually are. Strong emotion is rarely a feature of psychiatric or drug-induced hallucinations. The most significant problem with this hallucinatory theory of dreams is its requirement that the dream be considered a phenomenon independent of the rest of cognitive experience. Most cognitive states are viewed in the context of overall CNS function. Waking hallucinations are considered a symptom of an underlying disorder and are not viewed as an independent cognitive state. Sleep is not viewed as a cognitively independent phenomenon. If we consider sleep as a state independent of waking conscious, it is a state of unconsciousness or coma, a state of perceptual dislocation. In other words, sleep could also be considered a hallucinatory state in which actual perceptions (external objects) are negated by the CNS perceptual system.

There are advantages to addressing the hallucinatory quality of dream. If the dream is viewed as hallucinatory and delusional, dreaming can be considered as a valid model for psychosis. This perceptive of dream is supported by reports of disordered dreaming and recurrent nightmares in patients with some forms of psychiatric illness. This association as well as the correlation between dreaming and creativity will be discussed in Chapter 16. The imagery of dream can combine perceptual experiences in novel and creative patterns. Viewing dream as a hallucinatory experience does fit with shamanistic perspectives. Both induced hallucinations and dreams are used to obtain spiritual insight within the context of initiation rites. In this practice, the drug-induced hallucination is viewed to have the dream-like potential to assess components of mind not normally assessable to waking experience (Eliade, 1964).

Unfortunately, the primary use of the hallucination theory of dreaming has been in supporting simplified theories of dreaming, particularly activation–synthesis.

If a dream is merely a perceptual hallucination, dreaming can be better considered as a simply meaningless perceptual state based on primitive brain stem activity (REM sleep) of the self-referenced mind (Hobson, 1988). If the dream is as simple a process as a perceptual hallucination, dreaming can be postulated to be one of the processes utilized by the CNS during sleep to detoxify the system of unwanted memories of potentially pathological nature such as obsessions, hallucinations, and delusions. The hallucination theory of dreaming has been so utilized in supporting the "erasure" theory of Crick and Mitchinson (1983) who state their hypothesis as: "We dream in order to forget."

There are few who argue that a dream is a simple perceptual hallucination. Dreaming is not just a series of visual images produced in the state of sleep without perceptual input. Dreams also include emotions, memories, and thoughts set within the formal narrative structure of story. In the experience of dreaming during sleep, the dreamer often believes these experiences to be real. On waking these experiences are reported as dream. The dream experience can potentially be considered hallucinatory during dream. But in waking, the dream, no longer believed to be real, is a dream. Dreams are more than a hallucination. They are our memory of the cognitive functioning of our brain in sleep. Dreaming is a cognitive state with a neurochemistry, neuroanatomy, and electrophysiology at least as complex as waking.

CHAPTER 15

Dreaming and Storytelling

Something
Ought to be written about how this affects
You when you write poetry:
The extreme austerity of an almost empty mind
Colliding with the lush, Rousseau-like foliage of its desire to
* communicate*
Something between breaths, if only for the sake
Of others and their desire to understand you and desert you
For other centers of communication, so that understanding
May begin, and in doing so be undone.

      (John Ashbery (1985) *And Ut Pictura Poesis is Her Name*)

THE NARRATIVE STRUCTURE OF THOUGHT

Dreams are epics. The dream presents the episodic quality of an individual's life as it is lived, being closest in epic form to the soap opera – the continuing saga of our imaginary lives. A major difference between dreaming and waking thought is that in dreaming, experience is almost always presented in a narrative or story form:

"What then is a story?

A story describes a sequence of actions and experiences done or undergone by a certain number of people .... These people are presented either in situations that change or as reacting to such change. In turn, these changes reveal hidden aspects of the situation and the people involved and engender a new predicament .... The response to the new situation leads the story towards its conclusion ..." (Ricoeur, 1984a).

Most dreams are narratives occurring and often presented without applied organization, grammar, or expectation of critique. In the dream we can literally observe the "thinking of the body," and, with it, the birth of the literary process. Our dreams can be considered an exercise in pure storytelling whose end is nothing more (or less) than the organization of experience into set patterns that help to maintain order for the thinking system. The relevance of a dream's content to the dreamer's personal life may be less important than this function that dreaming serves as a structuring principle for life (States, 1993).

Narrative is a perceptual activity that organizes data into a special pattern representing and explaining experience. Narrative can be used as a way of organizing experience, drawing together aspects of spatial, temporal, and causal perception into story (Brannigan, 1992). Narrative organizes a group of experiences, sentences, or pictures so that they have a beginning, a middle, and an end. The beginning, middle, and end are not discrete elements such as sentences, rather they are relationships among the totality of elements that make up the story. Narrative becomes a way of globally interpreting a set of relationships involved in an experience or a sequence of actions (Brannigan, 1992).

There are typically narrative storylines to dreams. As dreams progress they begin to obtain a conceptual framework that include the preconditions for a narrative or story structure. Dreams are inherently self-organized in the structure of thought. The dreamer requires no training or critique to learn to present a dream as a narrative. The organization of the dream is dependent on the patterns of experience that each of us derive from the waking world. Dreams have a logical sequence of associations, a situational dynamic, in which each dream incidence occurs in response to the incidents that are already there in the dreamer's memory of waking experience. Dreams foreshorten and expand stories as is typical of waking narratives. Genuine narratives are structured in terms of a beginning, middle, and end, expanding or foreshortening temporal perspective in terms of plot requirements, reflecting and potentially varying a distinct narrative voice or point of view (Ricoeur, 1984b). Dreams, because they are intra-personal experiences, can disregard the requirements of communicability that is necessary for most stories (this is perhaps the major distinction between dream and fiction).

Dreams differ from the typical stories of fiction in that they have no beginnings or endings but rather seem to be all middle with occasional crises (States, 1993). Dreams are in constant motion, unstable, subject to instant revision and expansion. Dreams may not even have plots. It is as if dreams are trying to become genuine stories but typically fall a bit short (Hunt, 1991). Yet dreams are often organized in literary forms. These structural forms can be classified as dramatic, epic, or lyric. That these modes of story organization are characteristic of both dreaming and literature implies that these structural forms of organizing dreams may be essential combinational strategies, or ways of bracketing the world. When dreams tell stories, these stories are scripts made out of universal concerns, much as in literature. There are a series of consistent almost universally experienced dreams of flying, chase/attack, drowning, poor test performance, nakedness, and dreams of being trapped (Garfield, 1991). Plot in literary fiction is a continually evolving pattern of imagery and events (States, 1997). These same structures utilized to organize waking thought, may be the only organizational structures available to organize experience for the order prone mind (States, 1994). The narrative of dream reflects our attempts at comprehension of our world. It may be that this reflects a basic tenet of what is required of us in being human, "learning to understand and to be able to tell stories" (Foulkes, 1985).

That these modes of story organization are characteristic of both dreaming and literature implies that these may be essential combinational strategies, or ways of bracketing the world. Studying those narrative structures can lead us not only to understandings of the structure of the dream state, but also to understandings of how the mind organizes waking perceptions and memories. One of the purposes of dreaming may be to connect an individual's particularized knowledge with general knowledge in a way that waking experience will not permit. Bert States makes the case that dreams are epics representing the episodic quality of an individual's life as it is lived. In this way, our dreams imitate not only the content, but also the form of our waking experience. For the writer, focused on the process of storytelling, dreams can supply an invaluable and nightly renewable source of stories.

Dreams and fictions are both processes for connecting previously unconnected matrices of experience (Koestler, 1969). Both recreate human experience in a narrative form dealing in hypothetical or imaginary events (States, 1994). Narrative can be used to suture the viewer into the process of the fiction, pulling the viewer into the story based on the fear of being cut off from the narrative (Silverman, 1983). Storytelling differs from dream in that it is modulated by different constraints, primarily a responsibility to the reader who requires coherence, tension, and crisis.

THE MYTHIC STRUCTURE OF STORY

One way of looking at fictions is as waking dreams designed for other people. A piece of literature can stand in the place of a patient's dream or associational reports from the consultation room. Sigmund Freud had a deep interest in literature. His contribution to literature, however, comes not from what he has to say about literature, but from what he has to say about the human mind (Mahoney, 1987). As was suggested by Freud, many of us view a dream as a wish-fulfilling fantasy. Dreams can secure the apparent satisfaction of unrealizable desires. By imaging our desires satisfied, we can achieve a state of pleasure that simulates our ideas actually being satisfied. In cinema, it is suggested that we are offered this projective illusion in a manner akin to the psychic force of dream (Metz, 1982).

Story situations, master plots, are told over and over, often persisting through cultural change. Analysis of the narrative structure of fictions indicates that the structures of most are surprisingly consistent: the hero, the goal or objective, the helper, and the opposing force or hinderer. These can be considered the life and death forces of all narrative (Todorov, 1981). The repeating characters of world myth such as the young hero, the wise old man or woman, the shape-shifter, and the shadowy antagonist have psychoanalytic correlates with the figures that appear repeatedly in our dreams and fantasies. Dreams like fictions are made out of universal concerns and situations. When the scriptwriter uses such psychoanalytic structures and characters in story, they give the story the ring of psychological truth (Vogler, 1992).

Jung incorporated these figures, calling them archetypes, in his theories of personality and self. In dream and film interpretation this approach remains one of the most successfully applied, useful in describing basis, motive, and meaning of both waking and sleeping fictions. When the filmmaker incorporates such a mythic structure into a film, he or she is offered the possibility to evoke the peculiar kind of experience shared by all dreamers. The filmmaker can attempt to incorporate the dreamer's complete credulity for images with an autonomous power that appear without volition on the part of the dreamer. There is the potential through film to assess the power of a simple dream image to provoke the extreme emotion and the helplessness of the dreamer faced with his own creation (States, 1993).

Freudian approaches to criticism utilize the text to achieve a biographical sketch of the author. Jacques Lacan has extended post-Freudian approaches, utilizing psychoanalytic approaches to analyze the character in the story rather than the author. This approach tends to omit elements traditionally involved in literal analysis such as historical context, genre, style, ideology, and esthetics. The Lacanian focus is on the narrative text. Lacan focuses on the textual signifier of the narration itself outside any external conceptual bias that the reader or critic bring to reading for the world outside the text. For Lacan, the text embodies an organization of codes, signs, ideology, and structure that reflects our human interaction with the exterior universe (Kaplan, 1990). The structure of the writer's text is in this way "unconscious" filled with clues describing both the readers' and the writers' interaction with the world. This concept of text as an organization of language, codes, and signifying systems and its reader as an interpreter are now typical concepts in literature and film study (Kaplan, 1990). In many ways, the role of the critic and the role of analyst have become structurally the same.

DREAMS IN LITERATURE

Dreams are narratives full of affective associations that can be incorporated into waking dreams (fictions) and made available to others through storytelling in writing or film. Dreams often incorporate the same organizational structures as stories. They can provide the mythical structure that is often required of screenplay. In incorporating dreams, the writer can potentially incorporate and access for the reader the credulity and emotional power that typifies dreams. The writer can also assess the multiplicity of potential meanings inherent in dreaming by including dreams in text. The ambiguity of dream content was apparent even to Freud:

> To explain a thing means to trace it back to something already known, and there is at present time no established psychological knowledge under which we could subsume what the psychological explanation of dreams enables us to infer as a basis for their explanation (Freud, 1914).

Dreams are very different from waking stories. They are similar in that each narrative event in the dream places a limitation on the possible associations that may occur next (States, 1993, p. 53). In non-literate tribal societies with active oral traditions, dreams are a major new source for songs and poems (Tedlock, 1987). This use of dreams in literature persists today. Writers whose first drafts were derived directly from dreams include R. L. Stevenson, Joan Didion, Kafka, and H. P. Lovecraft (Hunt, 1991). Many authors will occasionally incorporate dreams into their literature and screenplays.

DREAMS IN SCREENPLAY

First time filmmakers seem especially prone to incorporating a dream into their screenplays. A review of the scripts accepted at Sundance Institute (Utah) film making labs from 1998 to 2000 shows that dreamscapes were present in 40–80%

of the accepted scripts from new filmmakers. A collateral finding was that in the scripts that eventually achieved production, these dream scenes were often deleted from the final version of the screenplay (Pagel et al., 1999).

Dreams can have different roles in the storyline of a script. The stature of a character that has dreams is often enhanced. That character can recover a mastery of events, "it was only a dream," or can bring back from the dream a moment of special insight (Porter, 1993). Shakespeare sometimes used this approach. In "Cymbeline" a dream provides a glimpse of Posthumus's inner world, validating the character's worth (Westlund, 1993). Shakespeare's use of dreams was often to provide affective meaning for the character. Dreams with themes of death, love, and money are found in "Hamlet," "Merchant of Venice," and others (Stockholder, 1993). By pointing at a character's imbalances, they often point out a route to attaining equilibrium. In this sense, the scripted dream explores imbalances and conflicts in the struggle for unity (Potter, 1990). Dreams can be utilized to display ambiguity, as in Hamlet where dreams are used to cast up images of doubt and indecision, without providing resolution. This is one of the ways that Shakespeare's Hamlet achieves its timeless character retaining its congruence with our modern condition (States, 1992).

The dream is a useful vehicle for the manipulation of time. Time can be expanded, contracted, or reorganized with flashbacks, flash-forwards, and dream inserts.

If used as an integrated element of memory, the dream can be used to induce expectations, anxiety, and desire in the viewer. If used skillfully, such a manipulation can provide insight and clarity. If used as an exposition to provide explanations, a dream sequence can call attention to itself as a devise and disrupt the dramatic flow of the story (Mehring, 1990).

THE LIMITS OF DREAM IMAGERY

Non-perceptual images are two-dimensional neurological pictures presented as a coded three-dimensional space. Imagery, thin as "thought," lacks some of the vitality and vivacity of waking visual perception being in comparison "thin," "dry," "two-dimensional," and "inert" (Scarry, 1995). As Aristotle states in De Anima, "images are like sensuous content except in they contain no matter" (Aristotle, trans. 1984). Compared to waking perceptual consciousness, there is something different, perhaps something missing in imagery. As the poet says, the inner eye, "Draws an outline, or a blueprint, Of what was just there: dead on the line" (Ashbery, 1979). Perceptually, the sensory is more vivid, detailed, and complicated than the imaginary realm. A certain level of perceptual acuity is required to ensure that we can infer the structure of an external material reality and safely make our way through it. Slant, reflectance, color, and illumination are key elements of perceptual world absent or indeterminate in fog. Fog in the physical universe approaches the condition of perceptions in the imagination. These visual limitations affecting imagination extend into the visual imagery of dream. This lack of solidity to dream images may be why we tend to view actual mist, actual gauze, filmy curtains, and blurry rain as dreamlike (Gibson, 1966). Sartre bemoaned the inertness of his volitionally developed daydream imagery:

> [the act of the imagination] is an incantation designed to produce the object of one's thought, the thing one desires, in a manner that one can take possession of it. In that act there is always

something of the imperious and the infantile, a refusal to take distance or difficulties into account. Thus the very young child acts upon the world from his bed by orders and entreaties. The objects obey these orders of consciousness: they appear (1991).

This may be why first time filmmakers are more likely to include dream images in their scripts, and why those images of the volitional imagination are more likely to end up on the cutting room floor.

THE POWER OF THE STORY

How do we integrate this realization that dream imagery is transparent and perceptually limited with those powerful, visceral experiences affecting all senses that can be a dream? The experience of flying in dream, the overwhelming fear of destruction in the nightmare, these are not thin, two-dimensional experiences. There are suggestions for answering this question that come from literature. Verbal art, especially narrative, consists of small black marks on a white page, with no acoustical or tactile features. Such a meager perceptual experience, however, has the potential to acquire the "vivacity" of an experienced perceptual experience (Scarry, 1995). The novelist or poet can recreate the experiences of climbing

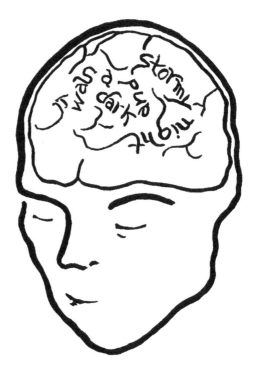

FIGURE 15.1 The writer accomplishes this magic not by description, but by creating for the reader the deep structures that gives rise to the perceptions that make the described experience look, sound, or feel a way that may seem more real than the reality of perceptual experience.

Everest, walking through woods on a snowy evening, or the intensity of love. A good writer can bring a reader to tears. How is this possible for a medium almost bereft of sensual content?

Elaine Scarry (1995) has analyzed the techniques that the great sensory writers, including Marcel Proust, Thomas Hardy, John Keats, and Seamus Heaney, use to incite the reader to experience mental images that resemble less the daydream and more the perception of actual experience. These writers do not achieve this goal by an intense description of a scene or experience. It is the scientist who is the expert at description. We send an individual into space, or perhaps into sleep, hooking up all the types of monitoring that modern medicine has developed for assessing physiologic functioning. We can obtain realms of data that describes the functioning of both body and brain – EEG, EMG, EKG, EOG, oxygen, and CO_2 levels, respiratory function and drive, body position, and if we wish data on neuron firing, ionic flux, glucose use, and metabolism in the brain. But we, the observers in the control room looking at all this data, do we have any personal sense of the person or experience that we are so busily describing? It is rather the writer who has the capacity to recreate another's experience. The writer accomplishes this magic not by description, but by creating for the reader the deep structures that gives rise to the perceptions that make the described experience look, sound, or feel the way that it did (Fig. 15.1).

The techniques that these writers use to recreate the intensity of perceptual experience often begin with a coaxing of the reader's imagination into a state of perceptual dislocation, antecedent to the material perceptions that are to be produced. For Proust this is the image of light or reflection passing across the walls of his room. For Hardy (1870) it is the colors created by the sun and shadows:

> Once more the cauldron of the sun
> Smears the bookcase with winy red,
> And here my page is, and there my bed,
> And the apple-tree shadows travel along.
> Soon their intangible track will be run.

Once the reader is coaxed into a state of imagination in which perceptions are limited and enfeebled compared to waking, interacting images are introduced. Images that are independently two-dimensional are able through their interaction to confer three-dimensional weight on one another. Once the reader is coaxed into this state of imagination, the writer presents a series of coherent steps for the reader to construct an image that has a vivacity, solidity, and persistence that is much like an image from waking perception. During this state of imagination, the reader gives up volitional control to the writer, and follows the writer's directions in recreating the perceptual experience. The writer brings the reader into a labor of imaginative construction. Here is an example of this approach taken from the opening paragraph or *Tess of the d'Ubervilles* by Hardy. Scarry (1995) has included what she calls "erased imperatives," the authorial directions for recreating the perceptions of experience:

> On an evening in the later part of May [picture this] a middle-aged man was walking homeward from Shaston to the village of Marlott, in the adjoining Vale of [hear the names] Blakemore or Blackmoor. [Look closely at the walker's legs.] The pair of legs that carried him were rickety, and there was a bias to his gait [picture the work of weight bearing and how his legs hold

his weight] which inclined him somewhat to the left of a straight line. [Create a geometric picture. Now let your eyes drift to his face.] He occasionally gave a smart nod, as if in confirma-
tion of some opinion, [drift now to the region of his skull] though he was not thinking of
anything in particular. [Look now at his arm and what you see.] An empty egg-basket was slung
on his arm. [Picture a second person.] Presently he was met by an elderly parson [look closely
at his legs] astride on a [look at the color] gray mare, who, as he rode, [listen] hummed a wandering tune. [Hear a voice saying] 'Good night t'ee,' [and look to see who it came from] said
the man with the basket.

The vivacity, solidity, and persistence of the perceptual world is reproduced for
the reader through this process of authorial direction beginning while the reader
is in an imaginary state.

A similar process may occur in dreaming. Dreams are visually experienced as
imagery that may be faint, fleeting, and two-dimensional. This cognitive process
occurs during a state of perceptual isolation free of new perceptual input, the
intrusion of waking thought, and outside of the volitional control of the dreamer.
These images are framed within a narrative storyline formally organized in ways
that are more associative and perhaps more bizarre than focused waking.
Thought that is more like the imaginative thought and the inner speech of creative process. Dream thoughts frame dream imagery in remarkable ways that create an entire perceptual universe seemingly real to the dreamer, and powerful
enough to disrupt sleep and affect waking attitudes and behaviors. As Harry Hunt
(1991) points out, "the creative tension required to produce novel, emergent
forms of self knowledge requires a staged collision between subjectivity and
objectivity that may characterize the dream" (1991). This process available to
every dreamer is what the novelist tries to recreate.

CHAPTER 16

Film: The Projected Dream

And if he were forcibly dragged up the steep and rugged ascent of the cave and not let go till he had been dragged out into the sunlight, would he not experience pain, and so struggle against this? And would he not, as soon as he emerged into the light, his eyes dazzled, be unable to see any of the things he was now told were unhidden?
No, at least not at first.

He would need, I believe, to first become accustomed to the light before he could see things in the upper world. First he would find it easier to look at shadows, next at the reflections of men and other objects in water, and later on the things themselves. After that he would find it easier to observe the sky at night, and the heavenly dome, and to look at the light of the moon and stars rather than at the sun and its light by day.

(Plato, *The Cave: The Third Stage (515 e 5–516 e 2): The Genuine Liberation of Man to the Primordial Light* – Part 1)

Film as a visual medium is often compared to dream. Even the first filmmakers had the capacity to create "dreamlike visages." During the first part of the last century, the Luminare brothers in their first attempts at moving pictures created a series of dreamscapes to demonstrate the possibilities of the new technology. Titled and numbered dreams, these foggy and jerky moving pictures resembled dream imagery more than waking perception.

Today's movie-makers still create dreams, but obvious differences are apparent when dreams are compared to film. Dreamers do not generally know that they are dreaming, while spectators know that they are at the cinema. Cinematic images are less personal and individual, and more social and ideological. For the spectator, the cinematic illusion may be more or less compelling than the perceptual illusion of the real dream. The projected dream of cinema is in some ways more formidable because it is a delusion experienced when awake (Cristie, 1994). The wealth of biographic and autobiographic reports from filmmakers describing a perceived association between dream and film imagery extends into the process of creation of the film.

MOVING PICTURES

At its point of origin there is a physical reality to film – an image on which a photograph is based. Of course, a photograph is a manufactured image of the world rather than an objective picture of reality. A two-dimensional photograph uses the techniques of visual perception, "the camera as eye," to produce an image that the trained viewer looks into to see an image. We have developed a way of looking at photographs different from the way that we look at paintings. We "look through" a photograph and at the object in a painting. Object-based attention, as is utilized in viewing a painting, focuses on elements belonging to the object while ignoring space. This neural process is one among mapping and is likely to use different cognitive systems than those used in "looking through" a photograph (Olson & Gettner, 2000). That experience of "looking through" is carried over into our experience of cinema.

In the hands of an experienced filmmaker a projected series of photographic images can be used to establish for the spectator what amounts to a complete disconnection with objective reality. For the spectator, it is often difficult, faced with the projected and screened illusion of the film, to visualize that the film was derived from an original reality of staged and directed photographs. The motion picture was a technical feat that preceded by many years any clear understanding of visual perceptual systems. Theories of perception present at the time of development of cinema were inadequate to explain the phenomenon of moving pictures. The technical background for development of cinema came from the interaction between a close-knit group of gentlemen scientists – Herschel, Faraday, Roget, Plateau, and Paris – and their shared involvement in the study of optical illusions and visual puzzles. They conducted their studies outside the focus of their main areas of work. Before the projection of moving pictures was possible, Daguerre and Niepce had to develop the photograph. Soon after the development of the still photograph, the actual moving celluloid of cinema was incorporated into a projection system. That invention has been variously ascribed to Thomas Edison's laboratory and to a photographer/inventor from Bath, England by the name of William Friese-Green.

A moving picture is a projected series of still photographs. When these images are projected at a speed of 16–24 frames per second, they are perceived as continuous. Beyond this threshold the periodic repetition of pictures dissolves perceptually into the continuity that is the motion picture. This effect exists because our process of observation is not instantaneous. Our visual system cannot process external information faster than it can transmit it internally. Beyond the perceptual threshold, periodicity in the incoming visual signal is read automatically and unconsciously as part of the film structure in which the message is encoded. We do not notice the gaps between pictures. In fact the gap between pictures can only catch our attention when it lasts long enough for us to perceive duration in it separately from the images that surround it. In early films, shot at 16 or 18 frames per second instead of the modern 24, we perceive a flicker based on their slow and uneven speeds (Chanan, 1996). Our visual perception in actuality requires even fewer images than the continuous projected photographs of the movie camera. We can recognize a two-dimensional visual depiction of an object that is presented in as little as 100 milliseconds (Klatsky & Lederman, 1993).

In cinema, the projection of images at a fixed rate creates a subordination of time to movement. In chronological time, the present is a point that moves continuously from the past into the future. There is a fundamental movement of time that has occurred and cannot be recalled. In cinema, time is measured dynamically, a process of action that defines the adjacent and contiguous spaces of the picture. Each shot is linked in sequence to the next, each sequence into parts and the parts into the whole of the film. An alternative reality is created that can be visualized outside the chronological reality of our external (non-film watching) lives (Rodowick, 1997). Our role as viewers is often "to work out" how our comprehension of the relevant film world is related to our normal modes of ordering and understanding perception in ordinary visual experience (Wilson, 1986).

Objects and forms are rarely visualized in the static condition since they are commonly in motion. Cohesive objects are perceived as capable of tracing only one potential path through space/time (Eilen et al., 1993). Our visual system anticipates the visual consequences of actions and events. Static images are projected along a potential trajectory before that sequence is ever shown. As early as the 16th century, painters discovered the power of intrapicture perspective in which the viewer follows the eye-line of the character. Our neural visual systems extrapolate progressions of actions to final results. Our attention often wanders and we base our memory of what we have seen on a succession of disconnected images (Kosslyn, 1994). Cinematographers and avant-garde directors, individuals pushing the capabilities of the medium to its limits in order to discover the possibilities, are among those who discovered the effects that alterations in object binding and motion could have on film and the viewer.

THE VISUAL LANGUAGE OF FILM

A visual language for film has developed on a trial and error basis over the almost 100-year history of filmmaking. Early works utilized a set camera acting as an audience for the performer. This early cinema had a freshness and energy of burlesque performance filled with a wild uninterrupted flow of gestures. Redirecting the camera from body shots to the close-up gaze of a character created a revolution in early cinema. Actors developed techniques of immobility and repression of gesture expressly for the new medium. The filmmaker and cinematographer could utilize the close-up shot to pull the spectator into emotional contact with a character. The carnivalesque, joyful, and dirty elements of early cinema were replaced with emphasis on interior emotions. Some feel that it was this close-up focus on the gaze, rather than minions of the devil, that brought desire, perversion, and obsession to the screen (Bonitzer, 1992).

Suspense could be introduced through the editing of parallel actions. Interspersing an image of a knife approaching a bared throat, against one of a car racing along a road in a cloud of dust brings the audience to a level of terror or anxiety over whether an anticipated rescue will occur (Bonitzer, 1992). Events that occurred over only a very short time such as the shower scene in Hitchcock's "Psycho" could be expanded with editing and slow-motion filming to develop a crescendo of suspense over an extended period of screen time (Zizek, 1992).

Master shots can be brought together in parallel and then into focus to bring together disparate tracks of the storyline. We have been trained as viewers to expect such establishing shots. Filmmakers can induce suspense by using close-ups where the viewer would expect such an establishing shot. Detail shots help to internalize psychological themes and structures for the viewer.

Film adheres to presence with everything taking place in the present tense. Flashbacks have fascinated both viewers and analysts. Formalism, structuralism, semiotics, theories of ideology, philosophies of memory and consciousness, as well as psychoanalysis are among the theoretical approaches that have been utilized in an attempt to understand the viewer's response to the flashback (Gabbard & Gabbard, 1999). In film study, a mix and match of such analytic approaches are often used in the study of flashbacks in order to allow the analytic techniques to comment on each other (Turim, 1989). This technique is very similar to that utilized by modern therapists in the content analysis of dreams.

Classical Hollywood narration rigidly subordinates the spatial and parameters of a shot to the causal logic of a story, ensuring a continuous line of narrative action. Transitions between shots are disguised encouraging medium unawareness and the experience of image and sound as if it were a continuous present. In order to maximize the spectator's experience of projective illusion, Hollywood films are designed to fit conventional story genres. Genres make it easier for the spectator to understand text by providing a framework of expectations based on other works in both literature and film that have explored the same area. Genres specify the expected content of narrative, drawing themes and images from other works in the same genre (Allen, 1995). The experience of projective illusion is in part a function of audience familiarity with the conventions of film narration, genre, and illusion.

The director of a film may subvert our expectations of narrative, genre, or character. However, it can be difficult to subvert the audience's expected illusion. Developed conventions of genre or character once learned by the viewer can neutralize that subversion. Clint Eastwood played with this expectation based on both genre and character, casting himself as the villain in *Unforgiven*. For many in the audience his character remained a hero, based on the actions of this character in other films, no matter what his actions were in the plot of this dark western.

VISUAL IMAGERY AND FILM

Cinematographic techniques imitate the way that each of us visually organize our perceptual experience. Film attempts to parody the perceptual/neurological movies that we create in our own minds. This approach can be quite powerful. The filmmaker is assessing the same perceptual system of waking and dream imagery that has the capacity to affect us emotionally and alter our perceptual view of the exterior universe. Hitchcock is purported to have said to Ernest Lehman on location for the filming of "North by Northwest":

> Ernie, do you realize what we're doing in this picture? The audience is like a giant organ that you and I are playing. At one moment we play *this* note on them and get *this* reaction, and

then we play *that* chord and they react *that* way. And someday we won't even have to make a movie – there'll be electrodes implanted in their brains, and we'll just press different buttons and they'll go "oooh" and "aaah" and we'll frighten them, and make them laugh. Won't that be wonderful? (Spoto, 1984).

APPARATUS THEORY

Freud developed his theories of the psyche in the same era that the techniques and process of cinema were being developed. Extensive crossover and borrowing took place from both directions. Early on, Freud realized that in cinema the spectator was conceptually part of the camera and screen. Part of the theoretical basis of psychotherapy was based on "apparatus" theory (Baudry, 1986). This construct has also been called the theory of "psychical locality":

> I propose simply to follow the suggestion that we should picture the instrument which carries out our mental functions as resembling a microscope or photographic apparatus, or something of the kind. On that basis, psychical locality will correspond to a point outside the apparatus at which one of the preliminary stages of an image comes into being. In the microscope and telescope, as we know, these occur in part at ideal points, regions in which no tangible components of the apparatus is situated ... Accordingly, we will picture the mental apparatus as a compound instrument, to the components of which we will give the name "agencies," or (for the sake of greater clarity) "systems." It is to be anticipated, in the next place, that these systems may perhaps stand in a regular spatial relation to one another, in the same kind of way in which the various systems of lens in a telescope are arranged one behind the other (Freud, 1914).

From this psychoanalytic perspective, the individual is not an onlooker to his own dream, rather he is enveloped by the dream just as a child is enveloped by his world. The individual is not fully present or in control of that visualized in dream. The subject does not see where a dream is leading, yet he or she follows (Lacan, 1979a).

Baudry extends this idea of the psychical apparatus to cinema. He suggests that this ability of cinema to envelope the spectator in the film process is why both a dream and a film experience can seem "more than real." Cinema can tap the ability of dream imagery to assess the unconscious mind. Filmmakers have been quick to pick up and apply this theory. Psychoanalysis has been used to provide a kind of map for cinema, what some have described as a "cognitive machine" (Kaplan, 1990). Apparatus theory purports to provide a cognitive map for the formulations of character, genre, and illusion that are used in filmmaking. Cinema includes both the transcendental subject and the object to be perceived by the viewing subject. It is as if there is an erasure of the material basis of the medium with the viewer attaining a temporal presence in the film (Allen, 1995). The film envelopes the subject much like the subject would be enveloped by his or her own dream. The subject is not in control, and is not able to see whether either film or dream will lead.

Jacques Lacan expanded Freud's apparatus theory to develop his own analytic model of the psyche. Of all the post-Freudian approaches to psychoanalysis, it is Lacan's approach that has proved to be most useful in film analysis. One of his most important psychoanalytic constructs is the Other. Lacan's Other is the ego-ideal of our self-concept that we see reflected in a mirror. This is the understanding that we receive of ourselves through projection. This understanding is dramatically ambivalent. It becomes our ideal self-image, however, it is always outside us. The concept of our mirror image self allows us to identify with the visual images that we encounter in the cinema, identifying alternatively with exhibitionist and the voyeur, the master and the slave, the victim and the victimizer. From Lacan's perspective, the appeal of cinema comes from tapping this interior psychic confusion between self and this projected Other (1979b). In the cinema, the spectator is in a situation where the apparent mirror of the screened film can bring the viewer to believe that he or she is present and involved in the production of the images on the screen (Kaplan, 1990).

Identification is the process in which the subject is displaced in his or her identification with the Other. Identification is a process with its own implicit ideology demanding sameness, necessitating similarity, and disallowing difference (Friedberg, 1990). Identification with a film star can extend outside of cinema to social and economic systems where identification with the star is used in entertainment, product, and political marketing. It is this process of identification that often leads to calls for censorship based on the concern that we or our children may imitate the harmful, illegal, or immoral actions of a character, actor, or star. Arguments for censorship are the converse of arguments for culturally supportive propaganda in cinema where identification is appropriately utilized to replace "wrong values" with socially or politically constructive "right values" (Friedberg, 1990). These psychoanalytic constructs of apparatus theory, mirroring, and identification can be used, at least in part to explain the power of film in personal, social, and political life.

FILM THEORY

Early on, serious debate took place as to whether film was actually art. Theorists suspected that all the filmmaker did was to create a record of what was put in front of the camera. Cinema might better be characterized as mechanically created entertainment. This entertainment was imitative of life rather than genuinely imaginative, and many considered it ridiculous to take film seriously (Passmore, 1991). Times have changed. The complex intellectual methodology of Freudian and post-Freudian film theory are the current "orthodoxy" of most university cinema study programs. This has taken place during the same period in which applied psychoanalysis has been in serious decline as a psychotherapeutic method for treating illness. Because of the many problems intrinsic to psychoanalysis as an approach to diagnosing and treating medical and psychiatric illness, Freudian-based film criticism has been viewed ambivalently by many of those skeptical of psychoanalysis. Yet Freudian and post-Freudian approaches to film criticism have proved useful, more useful than any other approach to film criticism that has been developed to this point. It is rare that a film critic uses psychoanalytic theory to make empiric statements about causality in a film; however, Freudian and post-Freudian perspectives are commonly used in attempts to facilitate an understanding of filmmaking, cinema, and the import of particular films. Filmmakers have developed complex methods of storytelling and have connected that process, intentionally or otherwise, with the anxieties of the audience. Filmmakers knowingly use psychoanalytic approaches in their films in attempts to induce in the viewer poorly understood and long-forgotten memories, and connect with powerful processes of primary thought (Gabbard & Gabbard, 1999). It can be argued that current psychoanalytic approaches are more useful in filmmaking and in the study of film than they are on the consultants couch.

Many film scholars use Freudian and post-Freudian methods to close-read individual films or to map the interior world of the director. Lacanian approaches focus less on psychoanalysis of the characters provided to us by the filmmaker and more on the imaginary and symbolic process of the viewer. This process has been called "suture" (Oudart, 1978). Suture is generally understood as a medically based metaphor implying that the gaps created in the process of editing are sewed shut to include the viewer (Dayan, 1976). The sutured viewer follows edited shots shifting from one character or scene to another by following the eye-lines (gaze) of the characters. The viewer accepts what is seen on film from multiple perspectives and directions as natural. In the classic example of shot/reverse shot editing, two characters are viewed alternatively over the other's shoulder. We do not ask "Who is watching?" because each shot answers the question of the previous shot (Gabbard & Gabbard, 1999). Dayan calls this the "tutor-code" of classical cinema. "Unable to see the workings of the code, the spectator is at its mercy. His imaginary is sealed into the film" (1976).

Researchers from outside the film genre, specialists in cognitive and neurological aspects of dreaming, have noted that cinematic language can be used to describe the phenomenology of dreams. Dream phenomenology is the way a dream appears and is constructed in the mind. When we watch a film we often incorporated images from that film into our dreaming (Cartwright et al., 1969). Dream scenes are often included as part of the storyline in film. It has been suggested that the incorporation of dreaming into film gives dream "bi-directionality" in which the interior world of dreams can be projected onto film. A filmmaker's dream displayed onto screen for viewing by an audience can have something to say that applies not just the individual who originally had the dream, but also to the problems of society (Bulkeley, 1996; Ullman & Limmer, 1988). These associations between film and dream have led some theorists to suggest that the fantastic creations of dream indicate that the brain has a fundamentally artistic component (Hunt, 1989; Pegge, 1962).

Film creates a cognitive state subject to outside control that is similar in many ways to dream. The filmatic interface can affect memory systems, visual imagery, and emotions. Film parodies brain-based components of the dream state presenting a visual storyline composed of associated images able to interact with the personal memories and emotions of the viewer. The film image like the experience of recalling a dream requires a feat of waking conscious translation and interpretation. The filmmaker can use the "bi-directionality" of dreams to affect the society and culture. Film has the potential to create an almost complete cognitive experience, fully outside the viewer's control, and using many of the same powerful cognitive processes utilized in dream.

Functional Uses for Dreaming

If dreaming has no function, as has been proposed by some theorists and researchers, it would be one of the only pervasively experienced physiological systems that has maintained existence in our species and others without having a function. Determining the functional significance of cognitive states is a problem throughout cognitive science, and logic protocols have been developed in an attempt to assess the functional significance of universally experienced cognitive states such as dreaming, sleep, emotion, and visual imagery (Fishbein, 1976, p. 8).

---

1. The characteristic has survival value at this point in development.
2. The characteristic has no survival value at this point, but is necessary to build another characteristic that will have high survival value.
3. The characteristic has no survival value at this point, but will eventually be combined with other characteristics, the combination eventually having high survival value.
4. The characteristic itself does not have high survival value, but it is the outcome of a characteristic with a dynamic function described in (1–3) serving as an indexing or indicating marker for that characteristic.
5. The characteristic itself does not have high survival value.

Class of Functional Significance.

Several of the central nervous system (CNS) physiological states that are associated with dreaming have questionable function. It has been particularly difficult to show clear functions for sleep. Each of us will spend a third of our lives sleeping. Sleep can become so important for each of us that we may put aside the most important events of our waking lives in order to get just a few minutes of sleep. It seems quite logical to suggest that sleep has a restorative function, restoring us for improved functioning on the next day. Yet it has been difficult to experimentally show that processes occur in sleep that can physiologically restore the organism. Deprived totally of sleep for days or weeks, an individual does not die. He or she experiences visual hallucinations and a decline in performance on boring repetitive tasks. Eventually the individual is unable to stay awake despite noxious stimuli designed to keep the person awake. The clear functional value that can be attributed to sleep is the improvement in waking performance that occurs when an individual has had sufficient sleep. Based on Fishbeinís scale, at our current level of knowledge, sleep has only a level 2 function.

Most of the functions that have been proposed for dreaming are psychological rather than physiological. The most generally accepted function for dreaming is in the resolution of emotional conflict (Chapter 11). Dreams have a narrative and visual structure that appears to be potentially useful for the writer and visual artists (Chapters 12 and 13). It has been suggested that dreaming allows us to assimilate contradictory models of the cognizant environment in a medium relatively free of logical constraints, for expanding and developing new information (McManus et al., 1993). Domhoff (1993) suggests that dreaming functions as a metaphorical attempt at problem solving. These functions proposed for dreaming are likely to have roles in creative process. Creativity is an important survival characteristic. It is potentially the definitive characteristic of our species that makes us different from most other animals. If dreaming does function in creativity, it can be persuasively argued that dreaming is at least a level 2 function – as important to the organism as sleep.

There has been little dream research focused on how dreams are utilized in waking behavior. Much of the work that has been done has come out of our laboratory based on interviews and a questionnaire (Chapter 19) that addresses dream and nightmare recall and the reported effects of dreaming on waking behaviors. The first modern study of the effects of dreaming on waking behavior (dream use) was done in 1991 in Kauai, Hawaii (Pagel & Vann, 1992). For this group of patients at a Family Practice medical clinic, increased reported dream use was associated with increased levels of dream and nightmare recall. Dream use was significantly higher in women and declined with age. In this diverse study population, equally ethnically divided between Caucasians, Hawaiians, Japanese, and Filipinos, no significant differences were found for dream and nightmare recall, or any of the dream use variables between these groups (Pagel & Vann, 1993). Several years later, a repeat study was done in the same population after the occurrence of a major pervasively experienced natural disaster (Hurricane Iniki). After this natural disaster, dreams were more likely to be associated with stress, but dream recall and incorporation into behavior did not change (Pagel et al., 1995).

When we began these studies, we were aware of anecdotal reports suggesting that dreaming had a role in creative process for both artists and scientists. Limited work has tried to address how writers, musicians, and artists use dreams in their work. We were lucky enough to be invited to work with filmmakers at the Filmmaking and Screenwriter Labs in Sundance, Utah (1994–1997) (Pagel & Broyles, 1996). Filmmaking is a creative area that has primary theoretic correlates to dreaming (Chapters 12, 15 and 16). Our work with filmmakers was extended on location during the filming of *Limbo* (1999) and *Silver City* (2005) both directed by John Sayles, with our results compared to studies of creativity in a sleep laboratory population (Pagel et al., 2003).

Dream effects on behavior were not limited to the creative aspects of these individuals' work. These studies indicate that creatively successful filmmakers, almost without exception, used dreams both in their work and lives at much higher level than our general population groups. The next chapters will focus on creativity as a cognitive process. We will look at attempts to define and characterize creativity. We will look at the association of creativity with dreaming, psychopathology associated with creativity, and the use of dreaming in the creative process of general population groups. Finally we will return to Sundance and look at how successful screenwriters, actors, and directors use dream in their creative process.

CHAPTER 17

Creativity

> *The apostle of affliction, he who threw*
> *Enchantment over passion, and from woe*
> *Wrung overwhelming eloquence, first drew*
> *The breath which made him wretched; yet he knew*
> *How to make madness beautiful, and cast*
> *O'er erring deeds and thoughts, a heavenly hue*
> *Of words.*
>
> (Lord Bryon – Childe Harold's Pilgrimage)

Creativity may be the definitive characteristic of our species. Yet, creativity is a descriptive term with multiple, poorly characterized definitions. The term is derived from the Latin base "creatus" – "to make or produce" or literally "to grow." The noun "creativeness" was first used in literature by Melville in *Moby Dick* (1851) to describe the mechanical "creativeness" of a lightening storm. In the 19th and early 20th century the study of creativity was called the study of "Genius" and more recently, the study of "Imagination." Webster (1996) defines creativity as artistic or intellectual inventiveness. Random House (1987) describes creativity as the ability to transcend traditional ideas, rules, patterns, relationships or the like, and to create meaningful new ideas, forms, methods, interpretations, etc. For the psychologist attempting to define a topic that can be measured, general definitions for creativity include: the process of bringing something new into being, or an ability to provide novel or unusual ideas (May, 1975; Guilford, 1965). Creativity can become an all-consuming process for an individual whose life is focused on discovering and preserving aspects of truth about life (Gardner, 1973).

Creativity is a descriptive term that is often used in more limited form in the context of the area in which it is being discussed. Creativity can be defined by the *environment* in which it comes about. Csikszentmihalyi defines creativity as the process by which a symbolic domain in the culture is changed (Csikszentmihalyi, 1996). It can be defined by the *product* of the creation. The artist often views creativity as the production of something that is both new and truly valuable (Rothenberg, 1990). It can be defined by the *process* of creating – the act of producing new information and testing whether it is objectively true (Garcia, 1991). For the visual scientist, the process of perceiving an image and cognitively participating forming a mental picture can be considered a creative act. For the theoretical neuroscientist the process of analog thinking can be viewed as creativity (Moliner, 1994). Creativity can also be defined by the *person* who is creative. This is psychological and intrapersonal creativity, when you have an idea that is fundamentally novel for you.

Creativity is often operationally defined as a process of problem solving, the process of intellectual inventiveness. Sternberg (1991) suggests that there are at least six facets to creative problem solving: (1) creative intelligence; (2) specific knowledge within the domain; (3) a certain style of mind; (4) certain aspects of personality; (5) motivation; and (6) a nurturing environment. Because creative problem solving has conceptual economic value in art, science, and other intellectual pursuits including politics, it has become a major focus for educators and cognitive neuroscientists. There are at least four major schools of approach.

THE GESTALT OF CREATIVITY

We can avoid trying to define creativity and creative process in an explicit, absolute way by considering creativity as part of a gestalt. "A creative moment can be described as part of a longer creative process, which in its turn is part of a creative life … The system regulates the activity and the creative acts regenerate the system. The creative life happens in a being that can continue to work in producing such creative moments" (Gruber, 1981). Creative and non-creative cognition are dependent on the same types of underlying cognitive processes with creativity an attribute of all human beings, who vary both its expression and its importance (Finke et al., 1992). "Creativity is founded on mechanisms … that can perhaps be approximated using the hardware and software of the machines we have today … Creativity is an automatic consequence of having the proper representations of concepts in a mind" (Hofstadter, 1985). Viewing creativity as part of a gestalt, visual thinking, spatial ability, pattern recognition, problem solving, and related forms of creativity are liked together along a continuum (West, 1991). The action of the child inventing a new game with his playmates; Einstein formulating a theory of relativity; the housewife devising a new sauce for the meat; a young author writing his first novel; all of these are creative, and there is no attempt to set them in some order of more or less creative (Rogers, 1954). Creativity can be defined as the "process of destroying one gestalt in favor of a better one" (Wertheimer, 1945), or "the intersection of two ideas for the first time" (Keep, 1957). Creativity is a feature not of the product of one's thinking but of the agent's cognitive act or process that resulted in the thought product (Dasgupta, 1996). From the gestalt perspective, creativity is a form of enlightenment, a process of transcending the mundane and routine in daily life. Creativity describes what occurs at that point, as when a mind perceives in disorder a deep new unity (Bronoxski, 1978).

THE CREATIVE PERSONALITY

There have been many attempts to describe the creative personality. Characteristics of the creative personality include, but are not limited to, fluency, novelty, flexibility, synthesizing ability, analyzing ability, reorganization or redefinition of existing ideas, complexity, and evaluation (Guilford, 1959). According to Rogers (1961) the internal conditions for constructive creativity are:

1. openness to experience including extensionality which is the opposite of psychological defensiveness, rigidity, and rigid boundaries;
2. an internal locus of evaluation in which the value of a creative person's product is established by oneself;
3. an ability to toy with elements and concepts, to juggle elements into impossible juxtapositions.

The external conditions for constructive creativity include personal freedom and psychological safety. These personality characteristics are broad and descriptive, qualitative and poorly amenable to quantitative testing. Psychological tests designed to access creativity often test for these personality characteristics. There is, however, only a limited degree of association between the levels of performance on these tests and an actual propensity to creative work.

PSYCHOANALYTIC CREATIVITY

From a "psychoanalytic" perspective creativity is a permanent operant variable of the personality. Freud stated that, [the] "source of creativity is the sublimation of energy into acceptable and fruitful channels" (1908). Freud believed that, "the creative imagination ... is quite unable of inventing anything; it can only combine components that are strange to one another" (1915). From the psychoanalytic perspective, in order to be creative, preconscious or unconscious material must be allowed to emerge (Bellak, 1958). Recognized instances of creativity are, in a sense, the "visible spectrum" of a far vaster range of manifestations of the creative unconscious mind (Harman & Rheingold, 1984). "It is a highly significant, though generally neglected, fact that those creations of the human mind which have borne preeminently the stamp of originality and greatness, have not come from within the region of consciousness. They have come from beyond consciousness, knocking at its door for admittance: they have flowed into it, sometimes slowly as if by seepage, but often with a burst of overwhelming power" (Tyrrell, 1946).

Carl Jung viewed creativity as the unconscious's conscious representation. He viewed creativity as a basic aspect even of non-creative individuals. "But what can a man create if he doesn't happen to be a poet? ... If you have nothing at all to create, then perhaps you create yourself" (Jung, 1965). Jung viewed himself as a creative personality incorporating in part Freud's negative view of creativity. "The daimon of Creativity has ruthlessly had its way with me" (Jung, 1923). Creativity is often suggested to be an individual's attempt to compensate for personality limitations or childhood abuse. Based on this perspective, creativity can be an attempt to reduce tension that may not be perceived consciously (Getzels and Csikszentmihalyi, 1976). Creative individuals are presumed in some circles to be compensating for hidden childhood trauma (Miller, 1988). Although such perspectives are common, they may reflect an over-reading of Freud's theories. Freud admired creativity. The types of childhood trauma that he postulated to lead to creativity were not necessarily abuse, but the kind of childhood traumas that we all have experienced (bereavements, separations, aggressions, and among others). Through creative expression, Freud suggested that we have the possibility to overcome childhood traumas. Freud suggested that overcoming trauma with a production that may have artistic value was to symbolically recreate the initial experience of distress (Mannoni, 1999). Creations, works of art, are the imaginary satisfactions of unconscious wishes, just as dreams are (Freud, 1925).

CREATIVITY AS DIVERGENT THINKING

Creative problem solving is often based on one's ability to go off in different directions when faced with a problem (Guilford, 1959). From a "self-expressive" perspective, creativity has been defined as "the ability to think in uncharted waters without influence from conventions set up by past practices" (Lee, 1940), and "the process of change, of development, of evolution, in the organization of subjective life" (Ghiselin, 1952). Creative thinking is often a search for meaning, which encompasses rapid bursts of ideas embedded in the sustained thought activities of the thinker (John-Steiner, 1985). Making variations on a theme can be considered as the crux of creativity (Hofstadter, 1985). George Bernard Shaw in his play Back to Methuselah gives this line to the serpent in the Garden of Eden, "You see things; and you say 'Why?' But I dream things that never were: and say 'Why not?'" Many of the tests that educators use to access creative capacity are designed to test an individual's capacity for divergent thinking (Gardner, 1973).

THE DISORDER OF CREATIVITY

Creative individuals are often not the most stable members of society. Writers, artists, and composers have higher levels of psychiatric illness than the general population. Diagnoses of depression and manic-depressive illness are common among professionally creative artists (Jamison, 1993). Studies addressing the psychopathology of major English poets have been particularly distressing, with 55% having a history of suicide, alcoholism, or mental breakdowns. Fully 18% committed suicide (Martindale, 1990). These studies are open to criticism. They are retrospective without control groups. Diagnoses are based on historical records and a reading of the artists work. Psychiatric illness is extraordinarily common, with some studies showing that 49% of the general population will have symptoms characteristic of major psychiatric illness in their lifetimes (Kessler, 1994). Major writers, poets, artists, and even scientists have provided us with detailed biographical and autobiographical information that can be retrospectively interpreted to reflect psychological illness. This level of insight is not available to us for other members of society. But the preponderance of biographical and autobiographical evidence does suggest that there is some association between creativity and psychopathology. As Lord Bryon once pointed out, "We of the craft are all crazy. Some are affected by gaiety, others by melancholy, but all are more or less touched."

Creativity can be considered a form of compulsive madness, driving artists by an urgent need to contact their dark, demonic and usually unconscious selves, or conversely an approach to promoting psychiatric health and even happiness (Kavaler-Alder, 2000). Whichever creativity turns out to be, depends on the particular individual involved in the creative process. Even Freud with his misgivings about the psychological origins of the creative process, admired the artist, and strove to develop his own creative skills. Consciously or unconsciously, creative individuals pursue both esthetic and psychological goals through their work. For some individuals creative work can be a path to cure, while in others a creative focus can lead to an exacerbation of psychiatric disturbance. Some theorists have proposed that a creative individual's sense of self can be repaired through using creative work to represent the self (Kligerman, 1980). Others view creativity in a more negative light – as an actual alternative to loving others (Gedo, 1983). It has been suggested that it is only when creative work brings the artist insight and resolution of psychic conflict can the artist begin to feel open to interpersonal relations. Many artists will never be able to achieve this capacity (Kavaler-Alder, 2000). Creativity can become entrapping for some individuals, with its preoccupation excluding interpersonal relations. Unhealthy artists can become addicted to the compulsive narcissistic reflection available in their art (Kubie, 1958). Although madness may be an inspiration, emotional and mental illness is more often a major hindrance to creativity for individuals working in creative fields (Rothenberg, 1990). Most psychiatric disorders by definition entail some loss of the ability to function in society.

Creativity requires not just divergent thinking. It most often requires bull like persistence, deep knowledge of the field of interest, and an ability to achieve societal acceptance. Compared to "normal" individuals, artists, writers, and creative people in general are both psychologically "sicker" scoring higher on a wide variety of levels of psychopathology, and psychologically healthier showing quite elevated scores on measures of self-confidence and ego strength (Jamison, 1993). Self-confidence and ego strength may be required for success in creative pursuits, but there are other personality characteristics that typify artists. Some of those characteristics affect the recall, and the use of dreams and nightmares.

DREAMING CREATIVELY

Creativity and dreaming have much in common. Both are self-expressive experiences that diverge from the step-wise, perceptual consciousness in which we live our daily lives. Dreams are often interpreted and creativity viewed best when addressed as global, gestalt experiences. Intense dreaming and creativity are associated with similar personality types and psychiatric dysfunctions. Both have been the focus of psychoanalytic thought and are among the most individual of experiences. Each person has a unique representation of what process is dreaming and which process is creative. Creative process has traditionally been considered to include the production of new and original products. But pass-times such as golf, hunting, reading, relaxing, and even television watching can be considered as personally creative parts of a creative life. Dreams have historically been included as part of the paradigm for the creative process of problem solving. This process described by Wallas in 1926 has four parts: preparation, incubation, illumination, and verification (Figure 17.1). The dream is the illumination.

Step 1: Choose the right night – when you are not overtired or inebriated and you have 10 or 20 minutes to prepare before sleep and 10 minutes, at least, for recording in your journal the next morning.

Step 2: Keep a journal, and write down your day notes in the period before sleep will relax you, clear your mind, and orient you toward your journal.

Step 3: Incubation, discussion – address with your conscious mind the problem you need to resolve, reviewing causes, alternative solutions, your feelings, and possible secondary gain involved in resolving the problem.

Step 4: Incubation phase – write down a one line question or request in your journal that expresses as clearly as possible your desire to understand the dynamics of your predicament (make this statement as simple as possible).

Step 5: Focus – put your journal beside your bed, turn off the light and close your eyes. Focus your attention and concentrate on your question, repeating the phase in your mind until you fall asleep.

Step 6: Sleep – usually the easiest step.

Step 7: Record – all dream memories should be written down in your journal as soon as you wake, either during the night or the first thing in the morning.

FIGURE 17.1 Steps for utilizing dream incubation in creativity process (from Delaney, 1979).

CHAPTER 18

Non-dreaming/the Use of Dreams in Creativity

"Finally, I believe, he would be able to look directly at the sun itself, and gaze at it as it is in itself, without using reflections in water or any other medium."

"Necessarily."

"Later on he would come to the conclusion that it is the sun which produces the changing seasons and years and controls everything in the visible world, and that it was also at bottom responsible for what he and his fellow prisoners used to see in the cave."

"That's the next conclusion he would obviously reach."

"And when he remembered his first home, and what passed for wisdom there, and his fellow prisoners, don't you think he would feel himself fortunate on account of his change in circumstance, and be sorry for them?"

"Very much so."

"And if the cave-dwellers had established, down there in the cave, certain prizes and distinctions for those who were most keen-sighted in seeing the passing shadows, and who were best able to remember what came before, and after, and simultaneously with what, thus best able to predict future appearances in the shadow-world, will our released prisoner hanker after these prizes or envy this power or honor?"

(Plato, *The Cave – The Third Stage (515 e 5–516 e 2): The Genuine Liberation of Man to the Primordial Light* – Part 2)

Creativity, difficult to define, has no definitive psychological test useful in its assessment. Tests can assess intelligence or divergent thinking, personality characteristics sometimes associated with creativity, but creativity, varyingly defined, develops in dissimilar individuals, and like anything new, it is often a surprise.

Dreams have a long history of involvement in the creative process. There are the autobiographical stories of Descartes' discovery of the scientific method, Kekule's discovery of the benzene ring, and Coleridge's writing of Kubla Khan – all purported to have come from dream. However, there have been few attempts to scientifically explore the potential association between dreaming and creative process. Dreams like creativity have a history of intrapersonal, psychotherapeutic focus. Until recently, there had been no population-based studies addressing perspectives toward dreams during waking. Dreams were special, meant to be assessed in a one-on-one basis by the trained psychoanalyst or electrophysiologist. In our early studies of dream use, questions were raised as to whether individuals in general population studies were really qualified to answer questions about their dreams. We were advised to be careful what we make of their answers. One answer that our subjects kept repeating: most were using their dreams in the waking process that they were calling creativity (Pagel & Vann, 1992).

CREATIVITY IN THE SLEEP LABORATORY

Creativity is most often studied in creatively successful individuals such as in the actors, screenwriters, and directors who were part of our Sundance filmmaker studies. Creativity is rarely studied in more general groups such as our sleep center patients in Colorado. These individuals, complaining of disordered sleep affecting their daytime functioning, came to our sleep clinic to have an overnight sleep study (polysomnography). In our sleep laboratory as in most clinical laboratories, more than 85% of patients are being studied to determine if they stop breathing during sleep. Obstructive sleep apnea (OSA) is common affecting up to 24% of males, and 9% of women. Most are loud snorers, who wake many times each night after episodes of apnea. Because of the multiple arousals from sleep required to re-initiate breaths, these individuals get very little sustained sleep during the night. Many are sleepy during the day, and have at least a mild decline in their cognitive functioning occurring secondary to either recurrent oxygen deficiency during sleep or daytime sleepiness. The cognitive deficits in OSA patients include declines in short-term memory and deficits in frontal cortex executive functions (Bedard et al., 1991). Executive functions have as a neuroanatomic base, the frontal cortex. Our understanding of the frontal cortex and its functions are limited. The cognitive executive functions that we understand to be at least in part based on the frontal cortex include: organizing daily activities, planning for the future, accurate self awareness, context appropriate goal oriented behavior, flexibility in response to changing environmental contingencies, task persistence despite distraction, and finally, creative problem solving (Duffy & Campbell, 1994; Fuster, 1989). If creative problem solving is a frontal cortex executive function, high level functioning of the frontal cortex is likely to be required of individuals involved in creative pursuits. Because of the decline in frontal cortex functioning associated with sleep apnea, we did not expect our sleep lab population that included many sleep apnics to be an especially creative group.

We asked these individuals to state what they considered their creative process (they could check off none), and then to rate their level of involvement in that process. That involvement was to be rated on a scale from a hobby or interest on up to an income producing focus. Subjects were asked to rate how often they remembered their dreams, on a scale from never to always. For this group the average reported frequency of dream recall was approximately once a week. Nearly 7% of respondents said that they never remembered their dreams.

When participants were asked to identify what they considered to be their creative process or interest, the largest response categories were physical activity (23%), crafts (15%), and music (15%). Nineteen percent reported having no creative interest. Physical activities were the most commonly described creative outlets with sports, particularly golf, hiking, camping, and gardening reported most frequently. There were other more sedentary activities described as creative pursuits such as reading, and relaxing, with six individuals stating that watching television was their creative process. A 45-year-old nurse reported "tanning" as a creative outlet. We did get responses from individuals involved in traditionally creative pursuits. Of 424 respondents, 80 were involved in different crafts, 33 were artists, 78 rated music as their creative outlet, and 25 were writers.

Ninety-two individuals responded that they had no creative outlet. Because there were so many different types of creative outlets described, our subjects' responses were categorized into one of three groups. Either the respondent had no creative interest (19%), a traditional creative interest that involved the production of creative products – art, music, crafts, writing, or theater (42%), or a non-traditional creative interest which did not involve the production of a creative product, such as games, history, reading, church, or physical activity (39%) (Table 18.1) (Pagel & Kwiatkowski, 2003).

Male and female respondents differ in their type of creative interest. Women are more likely than men (52.7% versus 35.7%) to report a traditional creative interest. Men were more likely than women to have a non-traditional creative interest (45.8% versus 27.4%). Responses of "no creative interest" were nearly equal between men and women (18.5% versus 19.9%).

TABLE 18.1 Self-reported Types of "Primary Creative Outlet" (N = 424)

Experiential creative outlets (N = 187)

Archery	Basketball	Bowling	Camping	Cars	Child
Church	Construction	Electronics	Family	Fishing	Games
Gardening	Genealogy	Golf	Grandparent	Hiking	History
Horses	Housework	Hunting	Lobbyist	Motorcycle	Puzzles
Ranching	Reading	Religion	Remodeling	Resting	Running
Shooting	Soccer	Sports	Tanning	Tinkering	
Watching TV	Weightlifting	Yard work			

Traditional creative outlets (N = 204)

Art	Calligraphy	Coloring	Computer	Crafts
Crochet	Dancing	Doodling	Drawing	Music
Painting	Photography	Piano	Quilting	Sewing
Theater	Woodworking	Writing		

DREAM USE AND CREATIVE INTEREST

The level and types of dream use in waking behavior varied markedly between the group reporting no creative interest and the other groups reporting any creative interest even at the minimal level of a hobby or interest (Table 18.2) (Pagel & Kwiatkowski, 2003). The dream use variables that increased as level of creative interest increased were dream use in creativity and work, in organizing daily activities, and in play (Table 18.3) (Pagel & Kwiatkowski, 2003).

This study is challenged, as are many studies of creativity, by the ambiguity of the concept of creativity. This ambiguity is obviously not only a part of research literature, but also a component, as well, of our responses when we asked our sleep lab patients to identify their creative process. How would you answer this question? We suspect that the answer is fairly easy for artists, musicians, those doing crafts and writers who have a classically defined creative process that results in a creative product, and much more difficult for others. Creative process that does not involve the creation of any creative product is the creative process for many, especially for males. The product of creativity for these experientially creative individuals is likely to be an intrapersonal change in attitude, memory,

TABLE 18.2 Dream Use Variable Means Describing Dream Effects on Waking Behavior Varying from 1 (= Never) to 5 (= Always) by Comparing Between No Creative Interest and Groups Reporting Some Creative Interest

	No Creative Interest ($N = 58$)	Some Creative Interest ($N = 424$)	F-value
Creative activities	1.33	1.69	6.67**
Organizing daily activities	1.26	1.61	7.03**
Attitudes toward others	1.43	1.71	3.85**
Attitudes toward self	1.48	1.84	5.04*
Adapting to change and stress	1.55	1.91	4.59*
Changing your life	1.22	1.54	6.74**
Planning for the future	1.31	1.70	8.38**
Foretelling the future	1.43	1.90	10.04**

*$p < 0.05$; **$p < 0.01$.

TABLE 18.3 Among Those with Some Level of Creative Interest (Level = 1–5), There Were Significant Correlations Between Level and Dream Use for the Following Variables

Dream Effects	Correlation (r-value)
Creative activities	0.11*
Working activities	0.12*
Organizing daily activities	0.11*
Recreation	0.11*

*$p < 0.05$.

and experience derived from the creative experience rather than an external product. As Carl Jung pointed out, creativity is a personality characteristic shared by most individuals even those not generally considered "artistic." "If you have nothing at all to create, then perhaps you create yourself" (Jung, 1965).

The major finding of this study was that dream use increased as the level of interest in creative process increased. This dream incorporation into waking is not just into creative process but into many aspects of life (Table 18.2). This finding was present for all of the types of identified creative process. The traditional creative groups that have a creative product do show dream use at higher levels than the experientially creative and non-creative groups. Dream use increased with increasing levels of creative involvement. However, the major difference in dream use was found between those individuals with no creative outlet and those reporting the lowest reportable level of creative activity as a hobby or interest. This finding suggests that when an individual has a creative process, dreaming can be incorporated into that process. There have been many studies indicating that successful artists use their dreams in their creative processes (Pagel et al., 1999; Sands & Levin, 1997). This study indicated that dreams have a role in creativity when an individual has a creative process.

NON-DREAMERS

For 2 years in the sleep lab setting, we used a questionnaire asking our patients whether they remembered their dreams. Of the 598 patients completing the questionnaire, 499 patients (93%) reported dreaming. However, it was not unusual for individuals to reporting not dreaming with 35/598 (6.5%) reporting that they had never experienced having a dream or nightmare.

Patients with OSA were more likely to report non-dreaming than individuals without apnea. Reported non-dreaming was associated with a significant increase in the number of times in which patients stopped breathing during the night. A patients' use of psychoactive medications can be used as a loose marker for underlying psychiatric and/or neurological disease. It was an interesting finding in this study that the non-dreamers had a lower incidence of psychoactive drug use than individuals reporting dreaming. Perhaps this reflects medication effects (Chapter 5). But there remains the possibility that non-dreamers are less likely to have psychiatric illness than those beset with dreaming and creative "madness" (Chapters 14 and 18). Could it be possible that non-dreamers are saner than the rest of us who remember the occasional dream? Personality theory suggests that individuals reporting low levels of dream recall are likely to have thick borders (Hartmann, 1991). These non-dreamers with their thick personality borders may be less likely to admit to psychological problems or complain of how hard is to function in the concrete reality of our modern world. Of course, another postulate is that perhaps these individuals were dreaming, and on our questionnaire saying that they were not.

As a follow up, we attempted phone interviews with the individuals who had stated on questionnaires that they never dreamed. We could not find 23%. Of the individuals that we were able to contact 12% said that they had misspoke themselves and that they really did dream. The largest group reported that they used to dream, but had not done so recently: 18% had dreamed as children but not as adults, 18% had dreamed into adulthood but lost that capability, and 23% had not been dreaming when they had completed the questionnaire but had began dreaming after treatment of their sleep disorder (sleep apnea or depression). Only 6% continued to assert never ever having had a dream. This result suggested that true non-dreaming was very rare in our sleep lab population (0.38%) −1 of every 262 patients. In order to have enough patients to complete the study, we had to continue questionnaires and interviews for 5 more years.

The classic way to determine whether someone is dreaming is to wake the individual up from different stages of sleep during the night and ask whether the individual has been experiencing thoughts, imagery, or emotions. We decided to use this approach with our patients who reported by questionnaire and interview that they had not dreamed for at least 1 year. Whenever they woke or were waken in the process of their sleep studies in the sleep laboratory, they were asked by the technician whether they had been experiencing thoughts, imagery, or emotions. We woke 16 true non-dreamers (individuals asserting that they have never ever experienced a dream) multiple times and asked for reports of what could be dreaming. In 36 awakenings, no one has reported a dream (Table 18.4) (Pagel, 2003).

Twelve individuals were included in this study who had dreamed in the past but not experienced a dream in over a year. These individuals only very rarely remembered dreams. The average time since their last reported dream averaged 17 years before their visit to the sleep laboratory. However, in three of the awakenings from three different individuals, a dream was reported (Table 18.5) (Pagel, 2003). For two individuals that report was the sensation of something going on – that a dream had been occurring. The other report had both content and storyline. The dreamer found himself walking naked along the Left Bank in Paris. This individual assiduously asserted that although this thought was a bit unusual and had occurred during his sleep, this was definitely not a dream.

TABLE 18.4 Non-dreamers

Individuals who report never have experienced dreaming (N = 16)
   Sleep laboratory awakenings
   Total awakenings (36)
      Stages 1 and 2 (18)
      Stage REMS (15)
      Stages 3 and 4 (3)
      Dreams reported (0) 0%

TABLE 18.5 Rare-Dreamers – No Dream Recall for At Least 1 Year

Total (N = 12)
Sleep laboratory awakenings
Total awakenings (32)
   Stages 1 and 2 (22) (two dream reports (9%)
   Stage REMS (10) (one dream report**) (10%)
   Total dreams reported (3) (10.6%)

Do these non-dreamers have dreams? Dreaming can occur without recall. For example, there are behavioral correlates of dreaming in young children and animals (a child's suckling movements, a dog's kicking, or running), and with delayed recall of dreaming (dreams not initially remembered, then recalled a day later). For the individuals included in this study who had never experienced a dream, if dreaming does occur, it occurs without any cognitive or behavioral correlates. Conversely, the rare – dreamers included in the study did occasionally report dreams when awaken in the sleep laboratory, most commonly on awakenings from non-REM sleep. Perhaps, despite reporting a lack of dreaming on awakening these individuals are dreaming. Until recently our understanding of dreaming and its relationship to waking consciousness has suggested that conclusion. Dreaming is occurring, everyone dreams, but these individuals must have a problem with recall and the transfer of dream mentation to waking thought. That certainly appeared to be the case for our rare-dreamer who was wandering naked in his thoughts along the Seine. But we cannot refute the possibility that these individuals reporting no dreams are not dreaming.

We perused the information that we had concerning these individuals: their demographic characteristics, their medical history, and their creative interests for characteristics they shared that could be related to their lack of dreaming. The two glaring demographic characteristics present in our non-dreamers are a lower level of employment 20% as opposed to 80% employment in the rest of the lab population, and a much lower level of creative interest (0.6 on a 0–5 scale) as opposed to an average level of creative interest in the lab population of 2.6. By comparison, all of the rarely dreaming group have been employed (57% are retired) and the level of creative interest reported was approximately that reported by the total lab population.

The Creativity of Non-dreamers

Currently, it is unclear what is blocking these individuals from the experience of dreaming. Their inability to dream could be biologic, psychologic, or a combination of both. This study did suggest that there is a group of individuals without obvious brain trauma or illness who do not experience dreaming. This is a small study and as such should be cautiously used as a generalization. But this data fits with the data from our dream use studies done at Sundance. In individuals such as filmmakers with highly developed creative outlets, dreams are heavily utilized in waking behaviors and creative process. Individuals, who rarely dream, had creative outlets and interests on par with the rest of our study group. But for our non-dreamers, those individuals who have never ever experienced a dream, involvement in creative process may be unusual.

CHAPTER 19

Dream Use by Filmmakers

Dream (1) – "I was hired by Joe Dante to write a script called 'Assholes from Another Planet.' That's how I visualized it, capital letters forming out of the stars just like in a film out of a black horror genre. These two guys come down from space and get a job in the New York City motor vehicle registration bureaucracy."

Dream (2) (two nights later) – "I had a film noir dream. The streets were cobblestone and wetted down. Bigfoot was lost in the dark city. He wound up in an alley with faces starring at him."

Dream (3) – "There was this black man in Harlem who couldn't talk. I had one of those realizations, 'Oh I know what's wrong, he's from another planet."

"So I decided to mix the comedy and film noir scenes into what was a skit more than anything else."

(John Sayles inspiration for *Brother from Another Planet* – Pagel, Crowe & Sayles, 2003)

When we began the Sundance studies, we were aware of anecdotal reports suggesting that dreaming had a role in the creative aspects of filmmaking. Filmmaking, the process of creating a film, is an interactive process between individuals in creative roles (actors, screenwriters, and directors); technically trained professionals (cinematographers, producers, and editors); and their support crew. We began our project with a series of interviews with individuals successful in their filmmaking field focusing on the creative roles of acting, directing, and screenwriting.

The interviews obtained during the 1997 and 1998 Filmmaking and Screenwriter Labs at Sundance were followed by a mailed questionnaire sent to all the attendees. This dream use questionnaire includes a compendium of validated questions on waking activities and behaviors that are altered by dreaming (Fig. 19.1) (Pagel & Vann, 1992). These questions were extrapolated statistically from a much larger group of questions initially utilized in our Hawaii studies to assess both dream and nightmare recall and the reported effects of dreaming on a wide spectrum of waking behaviors.

Please answer using the following scale			(Circle your response)					
Never [1]	Rarely (monthly) [2]	Sometimes (weekly) [3]			Often (3 times a week) [4]		Always (nightly) [5]	
Do you recall dreams?			1	2	3	4	5	
Do you have nightmares?			1	2	3	4	5	
Do you tell your dreams to others?			1	2	3	4	5	
Do dreams affect your waking activities?			1	2	3	4	5	
Do your dreams affect your:								
Decision-making?			1	2	3	4	5	
Emotions?			1	2	3	4	5	
Creative activities?			1	2	3	4	5	
Attitudes toward others?			1	2	3	4	5	
Working activity?			1	2	3	4	5	
Attitudes toward self?			1	2	3	4	5	
Play (recreation)?			1	2	3	4	5	
Planning for the future?			1	2	3	4	5	
Relationships?			1	2	3	4	5	
Setting your personal goals?			1	2	3	4	5	
Selecting entertainment?			1	2	3	4	5	
Adapting to change and stress?			1	2	3	4	5	
Organizing daily activity?			1	2	3	4	5	
Do your dreams: change your life?			1	2	3	4	5	
Foretell future events?			1	2	3	4	5	

FIGURE 19.1 Dream effects questionnaire.

We compared the responses from these filmmakers to those from our general population studies. Filmmakers were remarkably different. They were more likely to have nightmares and to describe their dreams to others. They had significantly

higher levels of dream recall and much higher levels of dream use in all categories. Since filmmaking is a group endeavor requiring diverse roles, we were able
to compare the responses of individuals in working roles (support crew, drivers,
and assistants) to the responses from the technically trained professionals (cinematographers, producers, editors, executive producers, and script supervisors) and
from the individuals in creative roles (actors, screenwriters, and directors).
Individuals in creative roles reported significantly higher dream recall, dream use,
and nightmare recall compared to the working group. The responses from the
professionals were somewhere in-between. Overall dream use was significantly
higher for the creative group compared to the working group. Specific waking
behaviors reported at significantly higher levels for the creative grouping compared to the working group include:

1. Dream description to others
2. Dream effects on waking activity
3. Dream effects on decision-making
4. Dream effects on emotions
5. Dream effects on working activities
6. Dream effects on attitudes toward others
7. Dream effects on attitudes toward self
8. Dream use in play/recreation
9. Dream use in setting goals
10. Dream use in adapting to stress
11. Dreams that change their life (life transformation dreams)

This study demonstrated that these successful filmmakers use their dreams at high
levels. Dream use for all our filmmaking groups was much higher than the dream
use reported in our general population studies. The filmmakers in creative roles
were far more likely to use their dreams than those with working roles. With this
increased use of dreams there was an associated increase reported in dream recall
frequency. Overall, dream use in creative process was one of the highest responses
for all of the groups. However, the affects of dreaming on behavior were not limited to the creative aspects of these individuals' work. This incorporation of dreams
into waking behaviors by members of the creative group was global. Responses
were significantly higher for 12 of the 17 waking behaviors queried, indicating that
dreams were used at high levels in all aspects of these individuals' waking lives
(Pagel & Broyles, 1996; Pagel & Kwiatkowski, 2003; Pagel et al., 1999).

SCREENWRITERS AND THEIR USE OF DREAMS

We had enough individuals in primary creative roles (screenwriting, acting, and
directing) to be able to compare their responses based on role in order to determine
if the type of creative process affected the type of dream use. Screenwriters incorporated dreams into their work at an intermediate level between directors (lower)
and actors (higher). The highest individual dream use questionnaire response for
screenwriters was for dream use in decision-making. Screenwriters were significantly more likely than the directors to use their dreams in decision-making. This
finding was consistent with our interview reports. In creating their alternative

realities, successful screenwriters often turn to their dreams to make decisions and to find alternatives to the approaches they develop during waking.

Screenwriting is very much a craft. Many screenwriters are also novelists, or successful in other areas of writing, but the screenwriter as a writer is required to have a finely honed sense of visual imagery. Writers tend to use their dreams in typical ways. Some screenwriters incorporate remembered dreams, almost verbatim, into their writing. Examples include John Nichols use of dreams in "Milagro Beanfield War" and Spaulding Gray's use of dreams in "Swimming to Cambodia" (Epel, 1993). Dreams are sometimes at the center of the seminal ideas for a work. One example is the quote at the start of this chapter – John Sayles' series of dreams that put in place the idea for "Brother from Another Planet" (Pagel et al., 2003). This screenwriter/director's description of dream use in screenplay has become a chestnut example of creative inspiration for filmmakers.

Personal dreamscapes can inspire storylines and be coherently translated onto film. However, it was unusual for screenwriters to report the inspiration a successful project coming from dream or the direct incorporation of dream images into scripts. Many more screenwriters used the inspiration of dreaming as background for their storylines, and in their decision-making process. Most reports of dream use by writers, describe dreams presenting alternative insights at points in which the writer feels "blocked." An example would be Stephen King who found the eventual conclusion for "It" in dream. Other screenwriters such as Gloria Naylor (The Women of Brewster Place) report extensive dreaming during screenwriting concerning the characters about whom they are writing (Epel, 1993). In our Sundance studies, instances where dreams were used in developing plot or script included:

1. Kayo Hato (Floating World) notes that images that he uses in his writing may come from dreams – the ideas of ghosts, winds, and spirits. He states that it would be "pretty stupid, ridiculous," if he were to use dreams to develop plots or scripts.
2. Robert Quezada (co-writer on Floating World) remembers that he was sitting in bed, half asleep, when he bolted out of bed and wrote a series of Guatemalan stories.
3. Craig McKay (actor, screenwriter) sees scenes from the script that he has written in his dreams. He reports an incident occurring at least a month after he wrote a screenplay in which the opening came to him in dream, complete in every detail. He says, "The screenplay is the springboard. It's the outline and dreams come like visions, letting the scripts tell him what's needed."
4. Susan Shilliday (Thirty Something, Legends of the Fall) notes that she rarely gets images from dream for her scripts, but that she has a goal to "get screen-play images as powerful as dream images." "I once dreamed up a pilot, but I could get anyone interested."
5. Sherman Alexis (Smoke Signals, Indian Killer) has frequent nightmares that he often incorporates into his stories. "If I didn't have nightmares, I'm not sure what I would write about. Maybe it's growing up on the "res," Indians have lots of nightmares to work from."
6. Stewart Stern (Rebel Without a Cause, and many others) notes that the opening scene of "Summer Breezes – Winter Dreams" came from a recurrent dream. In his dream, he is flying as a passenger in a plane when the engines cutoff and everything becomes totally quiet. The plane descends silently into a city, almost brushing buildings with its wingtips, heading toward the ground. This

dream on which the scene is based is recurrent and has alternative outcomes, with the plane crashing about 2/3 of the time. A later scene in the same film is also from dream. The female lead ascends a long vertiginous escalator populated at the top by her dead mother and other dead relatives. She looks back to see her children behind her at the base of the ascending stairs.

SCRIPTWRITING: THE DREAM-LIKE PROCESS

Storytelling can be thought of as an altered state of waking consciousness. Writing is a bracketed state of focused awareness in which one's consciousness of immediate time and space is dramatically altered (States, 1993). At the Sundance film labs, screenwriters often reported such experiences:

1. Theodore Thomas (Baby Photo) noted, "Most of my creative work, especially my writing, occurs in a quasitrance or dreamlike state. That is why I have sometimes found it difficult to respond immediately to critiques of the work, because since the writing was not arrived at in a logical way, my assimilation of new input almost always requires intuitive processing to see if it fits."
2. Lynn Seifert (Cool Runnings) notes her best work occurs when her creative impulse and her work merge effortlessly and her writing becomes "like a dream process, a juxtaposing of imagery."
3. "My work and my writing are the same. I try to get outside myself. Film imagery is similar to dreams, sometimes the same components with different characters. Films are dreams. What we try to do, is to create archetypical dreams" (William Yellowrobe).

Writing is a dream-like process for many of these screenwriters. Some describe their best work as occurring during periods of relative disconnection from conscious perceptual reality. This alternative conscious state has proven itself to have productive value for these screenwriters.

INCUBATING A SCREENPLAY

The first written records of the use of dream incubation have come from the writings on tomb walls from early empire Egypt. Incubation was incorporated into the diagnostic process of early Greek physicians. Anthropologic research has shown dream incubation to be a component of shamanistic practice in cultures as diverse as Amazonian Indians, Asian nomads, Native Hawaiians, Australian Aborigines, and North American Iroquois Indians (Delaney, 1979; Tedlock, 1992). Hieroglyphic or written records of this practice are preserved in ancient Egyptian, Sumerian, Jewish, and Greek records. Portions of the Torah, the Bible, and the Koran utilize types of dream incubation in attempts to justify "true" dreams that occurred to the prophets, as opposed to those "non-true" or rather "non-significant" dreams experienced by the rest of the populace. With its long history of use in creative process, it is not surprising to find that dream incubation persists as an approach in active use for augmenting the creative process of writers (Fig. 17.1).

Screenwriters often report using dream incubation in developing and writing their scripts. Before sleep they mull over problems confronting them in their writing. Answers to these conundrums can sometimes come in the dreams of sleep, at sleep onset, or during periods of disconnection when awake. Again from our Sundance interviews:

1. "Dreams can be empowering. They make you feel closer, helping you see yourself and your material. Sometimes in that hypnotic state just before sleep, I'll solve a problem that I've been focusing on" (Lynn Seifert).
2. "I think of something before bed and work it out in sleep. Sometimes I have weird nightmares about showing a film of my script with different imagery than what I thought that I had done" (Kayo Hato).
3. "I have dreams that have changed my creative process often on particular projects and sometimes in an overall transformation of my role as a writer" (Melissa Painter).

Incubation as a creative process has a long an illustrious history. Its persistence as an intrapersonal approach for solving the creative problems of screenwriters argues well for its utility in the creative process of writing.

DREAM USE BY DIRECTORS

The director has the most important creative role in filmmaking. Yet the director's role may be the hardest to define. On the basest level, the director actualizes the screenwriter's script. But as anyone who has read a script can testify, directing is far more complicated. From a particular script, it is not an exaggeration to suggest that hundreds of different films could be made. The director must visualize the world of the film on a four-dimensional basis. Since most films are not shot in sequence, the consecutive pattern of the storyline must be maintained. Each shot must be visualized from a cover and at least two additional perspectives. In each shot the actors must be coached to fulfill expected roles. Cast, professionals, crew, and setting have to be coordinated in a complex dance that often includes hundreds of people, multiple locations that may be on different continents, and a production timeline typically months or years in duration. The director must have the ability to integrate the various creative and professional input required to make a completed film. The actor, the writer, the cinematographer, the composer, and the editor, all of the individuals involved in these roles have different types of creative process. The ability to integrate all of this creative input is the creative role of the director. The person assuming the role of film director is responsible for both integrating creative performance and attempting to get the best work out of all the individuals involved in the creative, professional, and working roles required for filmmaking.

The seminal directors Ingar Bergman, Claude LeLouche, and Robert Altmann have credited dreams with parts of their product (Delaney, 1979). In interviews, biographies, and autobiographies, Bergman, Fellini, and Kurosawa have focused on the dream-like nature of filming in their work as directors. Fellini with characteristic bravado says, "Dreams are the only reality" (Chandler, 1995). Bergman states, "I discovered that all my pictures were dreams" (Kinder, 1988). In his major work,

"Dream," Kurosawa emphasized the duality between his creations and the real world stating, "reality remains reality" (1982). John Sayles utilizes both the visceral impact and global experience of dream-derived sequences in his works. "When thinking in pictures, its important to consider whether a certain image will have the same meaning to someone else that it does to you ... You build and pick your images so that anybody can get into the story on some level so that maybe people are drawn in deeper than they thought they could or would want to go" (1987).

In the Sundance studies, the overall average dream use for directors was slightly less than for actors and screenwriters. The highest questionnaire response for directors was for dream use in adapting to change and stress. The dreams of directors give them feedback about the process of filmmaking from conception to production. Some directors reported utilizing dreams in incubation. Almost all of the successful directors interviewed in this study reported high levels of dream integration into their waking lives and creative process.

DREAM USE BY ACTORS

In the process of filmmaking, the actor has a difficult role. Publicly it is the actor who often receives the accolades for the finished film. But in the process of film-making, the actor is likely to have limited control of how the performance is presented, edited and used in the film. Personality studies suggest that individuals who are successful in assuming the actor's role exhibit the qualities of extraversion, intuition, feeling, and judgment (Moreno, 1974). Actors utilize the script, support by crew and professionals, direction, and their interaction with other actors to develop a character role, but many actors rely primarily on an acting arsenal based on intrapersonal capabilities. Many utilize the input of dream.

Dreams can contribute to characters developed for performance. This perspective has been emphasized heavily in schools of acting based on the "Stanislavsky" method. "There exists only one impulse for creative art, namely, the creative powers each one of us carries inside himself." "The stage is where you can achieve a complete merging of yourself and the character of your part" (Magarshack, 1961). Stanislavsky refers to the creative dreams of the actor's imagination as the "visual images of his inner eye." Actors may use input from dream extensively in submerging their egos to assume the identity of another (Gardner, 1973). For the film actor, this approach characterizes each performance, in which the actor assumes as nearly as possible the role both on camera and off of the character being played.

SUNDANCE INTERVIEWS: ACTORS

Chris Morley (Gas Food and Lodging, many others) has recurrent dreams of being a soldier in Eastern Europe with the police chasing him, and returning to his initial acting group in which he learned to act. He's an intense user of his dreams. "I keep my rules in my dreams. I have an experience, say a relationship and it follows my expectations and my dreams say, 'I told you so!' I think that it's a shame that most people don't use their dreams. I use my dreams at all levels, especially in relationships and decision making."

Victor Wong (Big Trouble in Little China, Kundun, Three Ninjas) likes to intro-
duce himself saying, "I'm not a shaman. I just play one in movies." His dreams
are closely tied with his interest in the "I Ching," which he consults regularly
when he is acting.

David Straithairn (City of Hope, River Wild, Limbo; Good Night and Good Luck)
works as both a stage and film actor. He reports that the role of an actor can be
affected by roles in dream-like scripts or in scripts that are developed from dream.
Acting in a play by Sam Shepard in which the storyline and pace were both developed from dream, unexpected meetings and unusual associations affected the
lives of cast and crew outside the theater. While playing in the role of Robert
Oppenheimer, he had an experience that he describes as a visitation. "Here was
this character in my dreams, with his own history, his own story. I felt as if I were
being visited by a disembodied, discomforted entity that had its own reality outside me. It affected my performance. I felt that I had another audience besides
the one in front of the camera and responsibility to tell his story in as real a way
as I could."

SUNDANCE QUESTIONNAIRE RESPONSES: ACTORS

Successful actors have extraordinarily high responses to the questions included in
the dream use questionnaire. Actors' responses are significantly higher than those
for both screenwriters and directors. The highest responses for actors are for
dream use in creative activity, dream use in adapting to stress, and dream effects
on emotions and attitudes toward self. These high response questions are also the
questions that are answered with the highest responses, at far lower levels, by the
general population. What differs in dream use by actors is not specific but general. Actors incorporate dreams at high levels into all facets of waking life. Actors
report using dreams in developing character roles, and sometimes report achieving a connection with dream visualizations in their acting roles. The role of the
actor in assuming new and believable roles for each performance requires the
actor to bring all his capabilities and resources to the table. Dreams evidently provide intrapersonal information not available elsewhere for the actor confronted
with the creative task of developing and assuming a new role.

SUMMARY

The Sundance studies indicate that filmmakers use dreams far more than is typical of the general population. This association is not exclusive. There are filmmakers who do not use information from their dreams. Both John Amiel (Director)
and Walter Berstein (Screenwriter) are very successful in their field despite reporting limited dream recall and use. Such responses were, however, rare in filmmakers. The low levels of dream recall and incorporation into waking we found
in some of our general population studies were rare among filmmakers (Chapter 18).
Successful filmmakers utilize their dreams in all facets of their waking behavior.
All of the filmmaking groups, including crew and professionals had higher levels
of dream incorporation into behavior than the general population. A component
of creative success in filmmaking may be role appropriate utilization of dream.

This use of dreams varies with differences in creative role. Dream use appears to differ based on an individual's type of creative process.

Dream Use in the Creative Process

Filmmaking as a creative process may have more similarities to dream than other creative fields of endeavor. The filmmaker produces a visual storyline composed of associated images able to interact with the personal memories and emotions of the viewer to create an almost complete cognitive experience, fully outside the viewer's control, and using cognitive processes utilized in dream. However, the creative roles required for filmmaking can be found outside the filmmaking field. Screenwriters are writers. Actors have much in common with dancers and singers who also use their body and expression as artistic instruments. The role of the director in integrating multiple skills and the artistic expression of others resembles the roles of creative scientists, architects, visual artists, and sculptors. Dream use is likely to be typical for many of the individuals in such creative roles, who like actors are pushed to the limits of their personal capabilities, utilizing all of their available resources in their quest to create something that is new, different, and original compared to anything that has been created before.

CHAPTER 20

Dreaming and Artists

What I dream of is an art of balance, of purity and serenity, free from disturbing and engrossing subject matter, which would be for every mental worker, that is, for the businessman as well as for then man of letters, like a palliative, a mental soother, something similar to a good armchair that does away with the strain of his physical fatigues.

(Matisse, 1978)

Are dreams art? The answer has much to do with your definition of art. Dictionary definitions are variants of Plato's view of art being "a particular activity oriented towards beauty and its objectification" (Plato, Republic). Plato differentiates art from imagination: "a man-made dream for waking eyes" (Plato, Sophist). Art can be neurobiologically defined using such a Platonic definition – art can be considered, "as being an extension the function of the visual brain in its search for essentials" (Zeki, 1999). Art can be defined by what it is not, "this activity has an independent function, that is, it is not a by-product of other activities, not a mediating vehicle of various ideologies, not a maid-servant of theology and religious belief, and not an articulation of communal self-consciousness, but independent of these (though perhaps expressive of some of them) it is a self reliant activity" (Heller & Feher, 1989). A work of art is generally considered an intentional esthetic object – an artifact (or human product) with esthetic function (Genette, 1997). This definition serves as a marker for debate among both artists and those who appreciate and view artistic productions. This argument crystallized around Duchamp's "ready made" urinal when it was put on display at the Metropolitan museum for Modern Art. As Adorno (1970) states in his Aesthetic Theory: "Everything about art has become problematic: its inner life, its relation to society, even its right to exist". It is hard to argue with Bert States (1997) "... dreams have as much in common with art as they have, in another way, with unmediated waking reality."

The creative process of the artist has similarities as well to dream. Like dreaming, creative process relies on associative rather than liner or focused thought. It is an associative process developing from the artist's original intention. In the midst of creation the artist relies on spontaneous imagery and insight changing and redefining the initial concept. Through the creative process, the same initial concept or artistic intention can result in a multiplicity of dissimilar creative products (Hunt, 1989).

The art-like structure of the dream suggests that the human central nervous system (CNS) has a fundamentally artistic structure (Hobson, 1988; Hunt, 1989; Pegge, 1962). States (1997) proposed that "... art and dreaming are, in principle, the same activity in that they involve the same or highly similar mental processes (the production of images and narratives based on human experience) and that making art is no more elective, or voluntary, for the human being than dreaming." The images of dreams come from our experience and are likely analogies, metaphors, or models of memory structures. Much the same can be said of the creative inspiration behind an artistic piece of work (States, 1997). "... dream and art, in all its varieties, are manifestations of the same biological need to convert experience into structure and that dreaming, in all likelihood, preceded art-making in historical priority" (States, 1997). The realized dream has much in common with the completed piece of art. The work of art and the dream are simultaneously an intentional sign and a thing created in a way as to express something. Dreams can appear to be trying to be a created thing evoking meanings that might otherwise go unrealized (Husain, 1990).

DREAM USE IN CREATIVITY: PSYCHOPATHOLOGY

> The true artist will let his wife starve, his children go barefoot, his mother drudge for his living at seventy, sooner than work at anything but his art (Shaw, Man & Superman, 1905/1957).

Creativity can be a form of compulsive madness. Some human beings consider the artistic life to be a worthwhile endeavor, and they make sacrifices and call upon others to make sacrifices so that they can live a life of creating art. Others spend their lives in the detailed scrutiny or analysis of that art (Passmore, 1991). Creative individuals pursue both esthetic and psychological goals through work that can be socially and economically rewarding. This journey, however, is along a path made uneasy by Freudian suggestions that self-sacrificing devotion to art may be nothing more than a circuitous route to fame, power, and sex (Passmore, 1991). Freud believed that the psychological origins of the creative process were in the childhood traumas that we all have experienced. Through creative expression, the artist symbolically recreates the initial experience of distress, acting out, and working through experiences of loss and trauma (La Capra, 2000). Working from his role as a psychotherapist, Freud suggested that for some individuals creative work can be a path to cure, while in others a creative focus can lead to an exacerbation of anxiety and psychiatric distress:

> An artist ... is oppressed by excessively powerful instinctual needs. He desires to win honor, power, wealth, fame and the love of women; but he lacks the means for achieving these satisfactions. Consequently, like any other unsatisfied man, he turns away from reality and transfers all his interest, and his libido too, to the willful constructions of his life of phantasy, whence the path might lead to neurosis (1917).

In our own study of non-dreamers we found that psychopathology was more common in dreamers than in non-dreamers. The studies of Ernest Hartmann have shown that frequent, intense dreamers share with artists a psychological characteristic that he calls "thin borders." Individuals with thin borders see most issues as undefined. Few components of their lives are black or white. For individuals with thin borders, few questions have answers that are purely right or exactly wrong. They rarely report being fully awake or asleep. Life comes in shades of gray. Individuals with thin borders are more likely to report both nightmares and life-changing dreams. Such individuals have positive characteristics of openness as well as potentially maladaptive tendencies to vulnerability and fantasy. They are more likely to report hallucinations, visions, and hallucinatory like experiences. Diagnoses of schizotypical and border-line personalities, and schizophreniform disorder are more common in individuals with this personality characteristic (Hartmann, 1991). Both artists and intense, frequent dreamers often have this personality characteristic of "thin borders."

The use of dreaming for creativity can be viewed as a psychologically risky, but potentially success producing behavior. There are many components to creative process extending beyond the inner psychology of creativity that incorporates dreaming. To be successful in a creative role requires social support and the readiness for social change, training in a field of expertise, and dogged persistence by the individual. While creativity can be psychologically rewarding to the individual, it can be risky, financially and psychologically debilitating if non-successful, and very stressful. Few of the successful filmmakers included in the Sundance studies would suggest that they have chosen an easy, non-stressful life style. One of the requirements for maintaining a level of success in a creative field is likely to be a level of psychological and social stability that can tolerate the vagaries and stress of creative failure, and still continue to create. From this perspective, successful artists may be more stable than others, consistent with Freud's suggestion that through creative expression we have the possibility to overcome childhood traumas (Mannoni, 1999). Individuals operating on shaky psychological ground might be advised to demonstrate more caution than others in integrating both dreaming and creativity into their waking lives.

CREATIVITY AND DREAMING

Information describing the role that dreaming plays in creativity is fragmented and dispersed across a spectrum of humanistic and scientific fields. Creativity and dreaming are not independent fields of study. We have to range far afield to find this information, crossing epistemic field barriers that are rarely penetrated. Information on the role of dreaming in creativity is to be found in fields as diverse as sleep medicine, psychoanalysis, dramatic arts, art history, visual physiology, electrophysiology, linguistics, cognitive neuroanatomy, philosophy, psychology, psychiatry, and sociology. This information is hidden within the masses of scientific, descriptive, and theoretical verbiage that accrues within each field of study. To that incomplete list we can add another area of knowledge – film study – the study of the artificially created dream.

In the last few chapters we have presented the data and logic supporting each of the following conclusions about the relationship of dreaming to creativity:

- Both are concepts with individually developed definitions – developed from personal experience.
- Both are self-expressive experiences.

- Both diverge from the step-wise, perceptually defined consciousness in which we live our daily lives.
- Dreams and creativity are often viewed as global, gestalt experiences.
- Dreams have historically been included as part of the paradigm for the creative process: preparation, incubation, illumination, and verification. The dream is the illumination.
- The creative process may incorporate concepts and specific information from dreaming.
- Both creativity and dreaming have been the focus of psychoanalytic thought.
- Intense dreaming, and creativity tend to be associated with the same personality characteristics and psychiatric disorders.
- Creative individuals and those with intense dream lives may have higher levels of psychological pathology.
- Dream incorporation into waking, like creativity may occur in the same neuroanatomical area (the frontal cortex) and integrate as an "executive function" information from many different areas of the brain and psyche.

Despite a long history of anecdotal association, the role of dreaming in the creative process has received minimal scientific focus. This book pulls together existing data from a variety of fields with the filmmaking studies to demonstrate the many levels of association between creativity and dreaming. This evidence suggests that dreaming functions in creative process. It may be useful to look again at how dreams are used in the artistic process of creating a film. In a film, dreamscapes appropriately used in the story add complexity and emotional resonance. If the director inappropriately uses a filmed montage that is supposed to be a dream, the dreamscape can confuse and disorient the audience, and break up the narrative flow. These realizations apply as well to life outside film. When used appropriately, the integration of dreaming into waking consciousness can add levels of meaning and complexity to existence. From this perspective, a life can be art.

Not everyone may need to examine or even be aware of his or her dreams. Creativity can be a difficult, non-productive path. But for those artists, scientists, writers, filmmakers, musicians, and all the others who achieve personal reward from their own creative process, those individuals who take the risky path of living through their creativity, these are the individuals who likely need to use their dreams. Individuals functioning in creative roles need to be aware of their dreams in order to be successful. Dreaming, creativity, and art are intertwined cognitive processes. Creativity, while it may result in difficulties for the individual, is an important survival characteristic for the species, and perhaps the definitive characteristic making us different from other animals. It can be persuasively argued that dreaming is at least a level 2 function (see figure in Section 6) and as important to the organism as sleep (Fishbein, 1976). In fact, one of the most important functions of sleep may be to allow us to dream.

BORDERS OF MIND

It is an extreme Monist postulate that creativity and art are body and brain-based cognitive processes. However, we live in an extreme Monist era in which neuroscientists can logically postulate that the neural process supporting breathing in sleep are part of the mind, and at the same time, postulate that dreaming, creativity, and art are meaningless brain-based results of neural network functioning (Hobson, 1988). In this book we have defined brain-based processes as those for which scientists have been able to demonstrate a neural basis, and defined mind-based processes as those that cannot be explained using current scientific methodologies. Based on this definition, Hobson and his cohorts are in part correct. The dream images derived from analogies, metaphors, or models of personal memories utilize brain-based visual imagery and memory processing systems to produce non-perceptual two-dimensional images presented neurologically as a coded three-dimensional space. The visual artistic product is constructed from brain-based images much as the dream is structured based on analogies, metaphors, or models from memory. The emotions involved and utilized in dreaming, creativity, and art also have clear neurological correlates in the CNS. Imagery, visual perspective, and emotion are components of dreaming, creativity, and art. These are cognitive processes with significant brain-based components that can be artificially created. But dreams are visually experienced as imagery that may be faint, fleeting, and two-dimensional, with non-perceptual imagery lacking the vitality and vivacity of waking visual perception. How then, do we integrate this realization with the powerful, visceral experiences affecting all senses that can be a dream, a creative realization, or a work of art? (Scarry, 1995)

There is still much that we do not understand about the neurological basis of the processes of dreaming, creativity, and art. Large components of these cognitive processes can still be considered to be mind. The neural correlates of thought and the process of thinking are poorly understood. Neurological correlates for thinking are currently lacking. Because of these limitations of current technology, there is little research focused on the process of thought. The studies that do exist focus on waking thought, rarely address the types of associative thought most commonly utilized in dream and creativity. The cognitive process of thought involved in the integration of dream and waking imagery into a dream or a work of art is a complex and poorly understood process.

Thought in dream, creativity, and art can be viewed as the process utilized to frame imagery within a narrative storyline. In each case, thought denotes the personal presence of the dreamer or the artist in the creative process. The process of thought provides what Scarry (1995) would call the authorial direction of the dream. As in literature, thought provides a framework for the limited, two-dimensional images of waking imagery and dream, bringing the dreamer and the artist into the process of recreating perceptual experiences in an imaginative reconstruction. Thought has the capacity to frame dream imagery in order to create entire perceptual universes seemingly real to the dreamer, and powerful enough to disrupt sleep and affect waking attitudes and behaviors. Creative associative thought has the capacity to frame imagery in a way that the vivacity, solidity, and persistence of the perceptual world can be recreated and even surpassed for the dreamer, the artist, and the viewer of art. Such an integration of imagery, such a creation of "meaning" from disparate unassociated memories, can be considered mind. When we try to understand what constitutes the work of art, the artist, and the artistic product, we are attempting to understand the part of creativity and art that are mind. As Arnheim (1974) said, "The work of art does not ask for meaning; it contains it."

Models of Mind and Brain

Philosophy is like the mother who gave birth to and endowed all other sciences. Therefore, one should not scorn her in her nakedness and poverty, but should hope rather, that part of her Don Quixote ideal will live on in her children so that they do not sink into philistinism.

(Einstein, 1932)

This section marks the completion of the concrete neuroscientific portion of this journey to the limits of the mind/brain interface. In this journey, we have stretched both the hard neuroscience and the softer cognitive sciences of dreaming to the limits of current knowledge, addressing not only what we know but also that which we do not clearly understand. We have delineated in our approach those components of cognitive function that we can explain and understand (brain–body), from the areas of cognition which we have been unable to elucidate and explain with the techniques of modern science (mind).

Theories are based on models whose explanatory power is based on an etiological story that describes the development and occurrence of a particular attribute (Hardcastle, 1996). Standard model theory has concentrated on the development of complete models that correspond to a complete constellation of facts (Langholm, 1988). It would seem that a fairly large model would be required to explain central nervous system (CNS) function, the most complex physiologic system that we have attempted to scientifically understand. There are two primary ways of creating such models in the attempt to solve the mysteries of mind and brain function in cognitive process. There are top-down inductive strategies as are typical of philosophy, cognitive psychology, and artificial intelligence research, and bottom-up deductive strategies that are more typically characteristic of the neurosciences (Churchland, 1986).

Top-down approaches, overall simplifications, idealizations, and approximations, are required in the first stage of theory making, in the attempt to find the simplified principles that have the potential to explain other complex operations of the CNS. As Crick (1979) has said, it is important to select what problems to solve first because one is always on the brink of being thrown into a panic because of the complexity of the nervous system. Such top-down logical theory models can be too abstract to match experimental empirical data (Hardcastle, 1996). Yet, without empirical support such theories can come to be based more on belief than evidence. Without evidence, grand simplified theories are nothing more than unsupported folk wisdom or myth (Churchland, 1986). As Ramon y Cajal (1909) warned at the dawn of scientific neuroanatomy, [there is] "... the invincible attraction of theories which simplify and unify seductively."

The historic record suggests that even the best scientific theories currently accepted will someday come to be rejected as false. Theories persist as long as they prove useful in explaining experimental results and in driving research in the field. Theories come and go as experimental studies delineate specific results and the knowledge in a field is developed and refined. At its foundation, a basic empiric adequacy is required of any scientific theory (Costa & French, 2003). As Popper has pointed out, the method of science is a method of conjectures and refutations in which one proposes explanatory laws and then seeks evidence to show that the hypothesized law is false (Popper, 1935).

In this presentation, we have used an empirical ground-up (deductive) approach rather than a top-down approach based on simplified grand theory, to arrive at this point of building models based on experimental scientific data. In the next few chapters we will review the status of current models of mind and brain attempting to determine whether there is empirical evidence supporting these models. We will try to avoid the deconstructive urge to break down theories because they are incorrect descriptions of reality. At the platform of an adolescent's developing theory of mind is the realization that it is impossible in principle to maintain any certainty-preserving bridge between the realms of subjective experience and objective truth (Chandler, 1988). In attempting to move beyond adolescence and take an adult perspective, we will identify the conditions under which the current theories of CNS neuroscience are supported by evidence. The simplified top-down principles of current models can work when applied on a limited basis. In developing an adapted model of mind/brain we will use what can be saved of the current theories. We will delineate areas of cognitive processing where there is no evidence to support current theories of CNS processing, while describing the conditions under which these same theories continue to be true.

CHAPTER 21

Models of Mind = Brain

One ought to know that on one hand pleasure, joy, laughter, and games, and on the other, grief, sorrow, discontent, and dissatisfaction arise only from the brain. It is especially by it that we think, comprehend, see, and hear, that we distinguish the ugly from the beautiful, the bad from the good, the agreeable from the disagreeable ...

(Hippocrates)

Everyone agrees that all knowledge and thought are somehow related to the brain, but many thinkers pay little attention to this fact: the word "brain" is not mentioned in the index to Gilbert Ryle's Concept of Mind (1949). When philosophers think about thinking, they mostly examine their own thoughts. When scientists think about brains they refer not to their own brains but to observations made by neuroscientists.

(Young, 1987)

According to the monist, the nexus of interconnected neurons that we call the brain can be shown to be responsible for all human cognitive functioning. All cognitive function and behavior occurs secondary to brain activity. Based on this model, both the body and mind can be described in one box (Fig. 21.1).

In our modern era, radioactive isotope scanning techniques can be utilized to demonstrate real-time central nervous system (CNS) functioning occurring in association with motor activity, vision, speech, and even emotions. Based on such work monist postulates seem perfectly reasonable. The problems have been in the details.

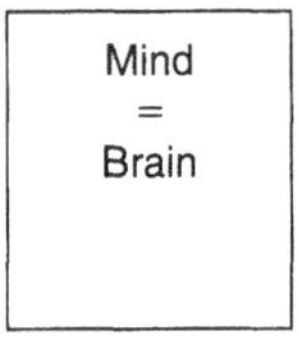

```
┌─────────────┐
│             │
│    Mind     │
│     =       │
│    Brain    │
│             │
│             │
└─────────────┘
```

FIGURE 21.1 Monist model of brain = mind.

TRANSMISSION-LINE NEURO-PROCESSING

Classic neuroanatomical theory is a bottom-up theory based on the transmission-line interactions of neurons communicating by means of electrical spike discharges across neurochemical synapses between axons and dendrites of the cells. This system of sequential neural interaction is extraordinarily complex. The human CNS is comprised of a hundred billion neurons, each with multiple synaptic connections and each with the capacity to respond to multiple neurotransmitters (Kandle, 2000; Schwartz, 2000). This system has powerful computational capabilities as well as the capacity to learn and develop neural maps based on recurrent patterns of neural electrical activation (Damasio, 1994). Computerized artificial intelligence (AI) systems using such a complex sequential system of interconnections have shown the ability to extend beyond the capacity of biological systems in perception, motor control, computation, and communication (Chapter 10). This system lies at the basis of modern neuroanatomy and neurochemistry and serves as the primary axis of CNS control of motor functions, perceptions, memories, emotions, and communication. We have been studying this system in detail since the dawn of modern medicine. Yet, we have yet to determine how this system accounts for higher CNS functions such as thought, perceptual integration, and creativity. Perhaps, all that is needed is more information, or the technical capacity to simultaneously record the activity of multiple, adjacent neurons (Churchland, 1986). Technology will lead to further and better understanding of the sequential transmission system, however it seems quite likely that this system of neural connections is but one of the systems active in the CNS. As we have pointed out in each approach that we have used to study dream, there is much that we do not know about CNS function. We have no idea how complex sequential neural interactions can lead to a state of consciousness. Sequential neural interaction theory is a bottom-up theory that has been extremely useful in explaining neural functioning at one-level neuron interaction. There is a point, however, in each of the neuroscientific fields of study where a limit is reached, explanations no longer hold, and we reach beyond the concrete limitations of transmission-line theory to that which we are calling mind.

AIM

The theory of activation–synthesis has been the most clearly stated and fully developed of monist top-down theories developed to account for brain function

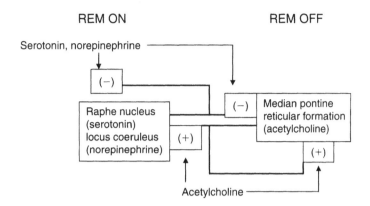

FIGURE 21.2 Activation–synthesis: the original reciprocal interaction model.

(McCarley & Hobson, 1975). Activation, in this theoretical construct, is "the aware-ness that is normal to an auto-activated brain–mind," while synthesis, "denotes the best possible fit of intrinsically inchoate data produced by the auto-activated brain–mind" (Hobson, 1988). Activation occurs in association with the primitive brain stem phenomena of rapid eye movement (REM) sleep. Synthesis occurs in the forebrain especially in the cortex and those sub-cortical areas associated with memory. In this theory, dreaming is an upper cerebral cognitive process utilizing the CNS activation associated with a primitive electrophysiologic state that we call REM sleep. The original version of this theory has been extended multiple times in the attempt to incorporate new experimental data on the neurochemistry of REM sleep (Fig. 21.2).

Allan Hobson and his cohorts at Harvard have extended this original activation–synthesis model of dreaming into a theoretical model that purports to describe all cognitive states. Extensive effort has been put into the development of this AIM model that proposes that all conscious states can be defined and dis-tinguished from one another by the values of three parameters: activation (A), input source (I), and modulation (M). Activation persists as a concept derived from the same laboratory's activation–synthesis hypothesis. This process is pos-tulated to include dynamic activation of the reticular formation, the PGO system, visual areas of the parietal operculum, as well as, the amygdale and paralimbic cortex (CNS areas involved in emotional processing). While these areas are acti-vated during REMS, a relative deactivation is present for the frontal cortex. Since there is apparently a neurochemical cholinergic triggering system for REMS in the brain stem, the system can be considered to have a cholinergic on–off switch. Input (I) can vary from the internal input of REMS dreaming to the external inputs of waking perception. Modulation (M) is considered to be a neurochemical process involving norepinephrine, serotonin, and cholinergic neurons. Modulation is postulated to reflect variations in behavioral states from deep sleep, to REMS, to waking thought, and even artistic performance (Hobson et al., 2003) (Fig. 21.3). Modulation has been a difficult concept for theorists that support AIM, as the number of modulating neurochemicals affecting REM sleep has recently increased in an exponential manner (Pace-Schott, 2003).

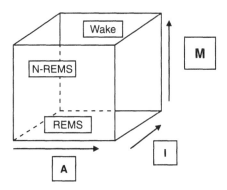

FIGURE 21.3 AIM model of neurocognitive states. Activation (A) is postulated to include dynamic activation of the reticular formation, the PGO system, visual areas of the parietal operculum, as well as, the amygdale and paralimbic cortex. Input (I) can vary from the internal input of REMS dreaming to the external inputs of waking perception. Modulation (M) is considered to be a neurochemical process involving norepinephrine, serotonin, and cholinergic neurons.

The AIM model has been adopted and extended by Francis Crick of DNA fame. He has proposed that activation of the reticular formation in the brain stem functions to focus attention generally in the CNS in order to organize neural operations. Crick proposed that rapid firing of thalamic cells indicates the activity of this system in the process of search and neural focus. He suggests that REM sleep is the primary model of the functioning of this system. He proposes that the results of the operation of this system are likely to the organization of neural nets in higher cortical regions (Crick, 1979; Churchland, 1986).

Both the AIM model and the search-attention model are attempts to extend activation–synthesis theory to other brain function. These theories suffer, however, from the same problems that have bedeviled activation–synthesis. Each theory relies heavily on the correlation of REMS with dreaming. It is the postulate of AIM that the neurons and neurochemicals that modulate REMS alter dreaming and other conscious states in a similar manner. The search-attention model relies on the example of REMS as the primary state during which dreaming occurs. None of the proponents of these theories have been willing to accept the possibility that REMS is not dreaming. Since REMS is but one of the brain states associated with dreaming, it is difficult to extend neurochemical, electrophysiological, or even pathological studies of the REMS state to cognitive states. Dreaming, imagery, thought, and waking consciousness are complex states poorly described by any of these theories. Activation–synthesis, AIM, and search-attention are at their origin simple theories relying on the correlation of REMS with dreaming. The complex overlay of these theories is built on the simple premise that REMS equals dreaming. If that premise is disproved, little empiric support for these theories remain, and the theoretic model for such top-down theories becomes better visualized as an inverted triangle with no support at its base (Fig. 21.4).

In the last 50 years, activation–synthesis, AIM, and search-attention have become the primary top-down monist theoretical models for mind and brain. These theories are based on a provably incorrect assumption. If in these muddied

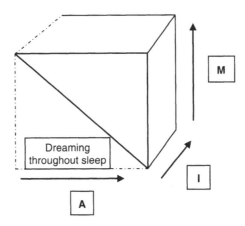

FIGURE 21.4 AIM model reflecting the lack of REMS = dreaming correlate.

waters we remove REMS = dreaming as evidence that mind = brain, what support remains for these monist theories of mind = brain? The state-of-the-art brain scans that demonstrate specific CNS neural activity occurring with REMS, can no longer be construed to demonstrate that CNS activity as associated with dreaming. The scans are demonstrating the CNS activity associated with an electrophysiological state we call REMS that has many physiological correlates that include increased dream recall relative to some of the other stages of sleep. These theories may be potential models for REM sleep, however, without REMS being dreaming, it is difficult to see how these top-down theories apply as grand theories of conscious states.

Other electrophysiological correlates for global cognitive states are not nearly as concrete as REMS = dreaming. Sleep onset is objectively a relative diagnosis, occurring in a gray area somewhere between waking drowsiness and the spindles of stage 2 sleep with the criteria of sleep onset varying from one individual to another. The behavioral criteria for perceptual isolation at sleep onset varies as well, with some individuals responding to noises and verbal cues well into EEG (electroencephalogram) defined sleep, and others having visual hallucinations that resemble dreaming in EEG defined drowsy wake. Removing the evidence of REMS = dreaming, brain death and seizures become the only clear electrophysiological correlates remaining that correlate brain activity with consciousness. Brain death has the medical-legal definition of an absence of detectable electrical EEG activity in the CNS. In our modern electrical era, brain death can be surprisingly difficult to demonstrate. In a setting such as the hospital ICU, DC and AC electrical fields bled from the plethora of probes and equipment utilized to keep patients alive leak into EEG recordings. In some ICU's respirators and monitoring equipment are disconnected and lights turned off in attempts to help the electroencephalographer be of legal and medical assistance in determining whether an individual has actually transcended from this existence. However, there is little question that an absence of EEG monitored electrical activity is associated with an absence of detectable cognitive behaviors. The behaviors associated with seizure activity in the CNS can be remarkably complex, particularly those associated with partial complex seizures, in which complicated motor activities, emotions, and dream-like cognitive experiences can occur. EEG correlates of most seizure activity can be mapped and traced to sites of origin in the CNS. The motor, perceptual, emotional, and even memory correlates of such events can often be site specifically traced to CNS activation occurring in association with the neural activity associated with the seizure.

Death and seizures, like taxes, are not nearly as philosophically satisfying as REMS = dreaming. If REMS is dreaming, a clear electrophysiological correlate can be postulated for a functional cognitive state. However, neither death nor seizures can be considered functional conscious states.

ALTERNATIVE MONIST THEORIES

Massively Parallel Neuro-Processing

The human CNS may be the most complex system that we have attempted to understand with our scientific tools and methods. The postulate that higher CNS functions are based on massively parallel neuro-processing evokes the science of complexity to bring us to the border between body and mind (Chapter 7). This bottom-up theory has been developed as an extension of sequential neural processing theory in the attempt to explain levels of cognitive processing such as analog thinking which appear to be beyond the capacity of computational systems designed on sequential processing. Artificial systems designed on this cognitive paradigm of highly distributed, parallel processing tend to be fault tolerant. Parallel processing neural net systems have applications in digital computation systems, addressing computational problems that are poorly amenable to linear mathematical approaches. Such systems add increasing levels of complexity to an already sequentially complex system. One current postulate is that at a certain level of complexity, a system is no longer just mechanical, biological, or cosmological. At that point, in order to interact with the external environment, the system self-organizes, takes on the attributes of mind, and becomes conscious. This hypothesis brings the sciences of complexity, AI, and quantum mechanics into the discussion of consciousness. The theoretical postulates based on these sciences are global and interesting, however, there is scanty current evidence for massively parallel processing as a basis for consciousness in either biological or artificial systems. This theory has the advantage of existing in a theoretic nether land where a lack of empirical proof competes with a lack of empirical disproof.

Physiologic CNS Electrical Rhythms

It is possible that the limits in top-down monist theories relate to their reliance on current neuroscience, and our current understandings of how the CNS works based on transmission-line neuroanatomy, neurochemistry, and neuropathology. In Chapter 6, we presented evidence that extracellular electrical fields are functional systems in the CNS. These extracellular electrical fields are known to affect the potential of transmission-line CNS neurons to fire. This system could function in setting cellular equilibriums, conveying information, supplying energy to the neurons, and potentially affect the developing CNS. This system may affect the expression of electrically sensitive and complex proteins such as DNA, affecting the expression of cellular memory that is stored in the genome (Pagel, 2005). This electrophysiologic system has the potential real-time capability of affecting gene expression.

This system is global, occurring diffusely in the CNS. The EEG rhythms are clearly associated with states of sleep and wake including focused waking (beta), drowsiness (alpha), sleep onset (sigma), and deep sleep (delta). These EEG rhythms may better characterize the stages of sleep than the classification methods in current use. Magnetic CNS scans attempting to define the developing origin of these physiologic rhythms demonstrate the rhythms evolving in typical patterns in generalized areas of the CNS (Rojas et al., 2000; Rojas et al., 2006). These rhythms vary between individuals. Abnormalities in this system can be seen in pediatric attention-deficit or hyperactivity disorder (AD/HD) and obstructive sleep apnea, as well as in adults with fibromyalgia, chronic pain and fatigue, and primary insomnia (Pagel et al., 2006). Psychoactive medications are known to affect these background EEG frequencies in consistent patterns based on the known behavioral effects of those medications (Chapter 5). Individuals with extensive training in meditation can induce changes in this system.

Our understanding of this system remains limited, and many neuroscientists would debate whether the physiological EEG rhythms are a functional CNS system. Other authors postulate that CNS systems with the ability to affect DNA expression are providing the underpinnings of consciousness (Rossi, 2002). It is clear that changes in background EEG frequency are associated with changes in global cognitive functioning. It is tempting to postulate that this electrophysiological system could be a biological correlate of the ill-defined cognitive processes that we are calling mind. Specific correlates between mind and CNS physiological rhythms remain speculative and like REMS = dreaming based more on belief than evidence.

PROBLEMS WITH MONISTS

We live in a Monist era where everyone seems to know that mind = brain. Clear brain-based correlates exist for perceptual processing, motor interfacing, communication, and computation. Developing evidence is strongly suggestive that much of the cognitive processing involved emotion and memory are based on brain-based neural processing. Limits of the Monist perspective have been discussed in detail. Attempts to correlate mind-based processing with specific markers of neuroanatomy, electrophysiology, or neural functioning have been notoriously unsuccessful. The REMS = dream correlate is a primary example of both the attraction and the limits of Monist belief systems. The global states, such as sleep and consciousness, are poorly described by current bottom-up neuroscience built on systems of sequential transmission-line neural connection. There are cognitive processes in the CNS including focused waking thought, intelligence, dream and associative thought, feelings, and creativity for which little evidence exists supporting a brain-based correlate. Cognitive processes for which we have clear brain-based CNS correlates have components that we cannot describe with current technology. Examples include the associative visual imagery of dreaming, and the feelings associated with emotions.

Artificial constructs of cognitive processing can approximate some of the CNS-based cognitive processes. AI has proven capacity beyond that of biological systems in perceptual and motor interfacing, communication, memory, and computation. Scientists have had less success constructing artificial systems capable of higher-level cognitive processes such as intelligence, thought, and creativity. There is little evidence that such systems have attained consciousness or patterns of functioning that we could call processes of mind. Such systems have not demonstrated capacities for focused thought, intelligence, associative thought, creativity, associative visual imagery, or feelings.

There are options that the true monist that can utilize to overcome this inability to find brain-based correlates for the cognitive processes of mind. The monist can define mind as those brain-based processes that we do understand. Here is an example of the logic of this approach:

1. We understand the brain-based origin, perceptual processing and motor presentation, as well as, components of the cognition (dreaming) involved in emotional processing.
2. Feelings constitute a mode of thinking, a style of metal processing, that integrates and addresses emotion.
3. Emotions are characteristic of dreaming.
4. Dreaming is surely a mind-based process.
5. This means that we understand the biological basis of feelings as an aspect of mind (Damasio, 1994).

This circular approach can also be applied to dreaming. The biological basis for emotions and visual imagery are well described, as are the specific neurochemistry and neuroanatomy involved in an electrophysiological state (REMS) that is associated with dreaming. Yet dreaming occurs outside REMS. The REMS = dreaming correlate indicates that if dreaming occurs, REMS must be occurring as well. Dreaming occurring outside REMS must then be occurring during a state that is an incomplete, not fully formed, version of REM sleep (Nielsen, 2003).

There are other approaches that can be used to emphasize the brain-based basis of dreaming. If dreaming is a state defined by the components of the state that utilize waking cognitive processing (emotion and visual imagery), and that is what a dream is, then we understand the biological basis of dreaming (Hobson, 1994). To fully explain dreaming as a brain-based process, the problem remains of the thought content of dream for which no clear brain-based correlate has been evident. By postulating that thought in dreaming is logically impaired and bizarre, thought content as a component of dreaming can be de-emphasized and denigrated. These approaches of re-definition and re-emphasis are used to support the authors' theoretical hypotheses while ignoring and de-emphasizing contrary evidence (Kahn & Hobson, 2005).

Monists sometimes react to such criticism by suggesting that future research based on newer and more specific technology will be able to delineate other of the processes that have been postulated to be mind. Brain scanning studies have a qualitative, pictorial power that often transcends scientific requirements of quantitative proof. Brain scanning studies have been interpreted to suggest that recently discovered brain sites are the basis for both spiritual and political beliefs. Such perspectives are amazingly Newtonian, suggesting that once we know the location and interactions of every neuron, we will know the inner workings of the mind. As we noted in the introduction, the cognitive sciences are quite young. The mature sciences such as physics, mathematics, and cosmology went through this same process a century or more ago. Concrete Newtonian approaches can still be used to explain much of those areas of science. In each of these sciences, such an approach failed when used in the attempt to theoretically explain areas for which the scientific evidence was contrary or inconsistent. What we actually know of the CNS is based on repeatable experimental evidence rather than global theory.

The alternative monist theories based on physiologic electrical rhythms and massively parallel neural processing are based on limited experimental evidence. There is increasing evidence that the physiological electrical rhythms are part of a functioning system in the CNS, however, the actual function and effects of this system remain debatable (Chapter 6). Massively parallel neural processing is currently a philosophical theory. These theories can be used to conceptually explain some of the limitations of classic sequential transmission-line neuro-processing. These are global processing theories, better able to explain such global states as sleep, thought, and consciousness. These systems have the capacity to involve cellular memory and control systems in cognition. These systems may offer the best hope for saving Monism by delineating aspects of the brain able to produce others of the cognitive processes that we are calling mind.

And still everyone seems to know that mind = brain. This may in part stem from "Cartesian anxiety," an unrequited yearning for some absolute foundation for certain knowledge (Bernstein, 1983). As John Searle (1984) bluntly states, "brains cause minds." Yet there is only limited evidence for the functioning of CNS-based neuro-processing systems in mind-based processes such as focused thought, intelligence, associative thought, feeling, creativity, and associative visual imagery. Monism is a theory requiring belief. And so many want to believe.

CHAPTER 22

Mind Not Brain: The Cartesian Psychoanalyst

*I should no longer fear lest those things that are daily shown me by my
senses are false; rather the hyperbolic doubts of the last few days ought to
be rejected as worthy of derision – especially the principal doubt
regarding sleep, which I did not distinguish from being awake. For
I now notice that a very great difference exists between these two;
dreams are never joined with all the other actions of life by the memory,
as is the case with those actions that occur when one is awake.*

(Descartes, 1641, p. 100)

From the purely Cartesian perspective, the cognitive processes called mind exists outside the biological processes of the brain. If a cognitive process is brain based it is not a process considered to be part of the mind (Fig. 22.1).

The Cartesian perspective can be based on the belief that the cognitive processes called mind come from outside the individual. The religious and spiritual overtones

FIGURE 22.1 Cartesian model of mind and brain.

of the Cartesian perspective persist into the current day, and are sometimes presented as its basic characteristic, "The attraction of dualism is that it seems to enable the mind to be a creative agent outside the ordinary causal nexus and perhaps account for various religious convictions about the spiritual element of humans" (Bechtel et al., 2001). Among psychoanalysts, Jung came closest to this perspective with his conception of shared archetypes, the shared universal symbolism of dreaming and art.

Alternatively, the Cartesian perspective can be based on the concept that mind-based processes are those that are not fully described by transmission line neural processing. It was Descartes' original postulate that some cognitive processes are brain and others mind. Even in our Monist era, this perspective has considerable resonance. It is the basis for the separation of medical from psychiatric diagnoses. If a disease has a biological basis, an abnormality of body functioning can be demonstrated. Evidence-based tests utilizing techniques of scanning, chemical assessment, tissue samples, or electrophysiology can be used to support the diagnosis. Psychiatric illnesses differ, generally occurring without obvious describable structural abnormalities and without biological markers to support or rule out the diagnosis. Psychiatric diagnoses are based on an analysis of behavior from a physician or another professional trained to recognize psychiatric forms of illness. Medical illnesses are diseases of body. Psychiatric illnesses are diseases of mind.

DISEASES OF THE MIND: THE PSYCHOANALYTIC APPROACH

Freud was trained as a neurologist and based on this training, generally supported the postulate that psychiatric illnesses have a biological basis. Freud was among the group of physicians that developed and attempted to validate psychoanalytic approaches for describing the behavior and presentation of patients with psychiatric illnesses. One of Freud's most important insights was that psychiatric illnesses could be analyzed without resorting to body-based tissue or scanning tests. This approach persists as a primary focus of current psychiatry. Specific psychiatric diagnoses are classified based on criteria of specific behavioral complaints. Particular diagnoses require that the individual meet selected and specific criteria for each diagnosis (DSM-IV, 1994). Since almost all psychiatric disorders result in disrupted sleep, these criteria generally include patterns of disordered sleep.

Freud and his adherents developed psychoanalytic approaches that could be utilized in diagnosing and treating individuals with psychiatric illnesses. The trained psychoanalyst could use these techniques to diagnose and treat abnormalities or limitations in cognitive processing and functioning of individuals suffering from psychiatric illness. The primary tools of these psychoanalytic approaches are free association and dream analysis. Freud utilized dreams as a window into understanding and analyzing the problems and processes of psychiatric illness (1917). Freud used the approach of free association of dream content to explore the associative patterns underlying dream content. With this technique, each line of dream-based associations could be followed into latent thoughts and memories with internal connections often quite different from the connections of the original dream plot (Hunt, 1989). An individual's psychic structure and dynamic could be inferred by the psychoanalyst from information derived from free association and the associative interpretation of dreams and utilized in developing a therapeutic plan for the treatment of psychiatric symptoms. "… the study of dreams is not only the best preparation for the study of the neuroses, but dreams are themselves a neurotic symptom, which, moreover, offers us the priceless advantage of occurring in all healthy people" (Freud, 1933). Psychoanalytic dream interpretation often focuses more on these associations of the reported dream than it does on the manifest dream report from the patient.

THE LIMITATIONS OF PSYCHOANALYSIS IN THE DIAGNOSIS AND TREATMENT OF PSYCHIATRIC ILLNESS

Freud viewed psychoanalysis as an attempt to extend the scientific method into the investigation of the mind (1933). For the first half of the 20th century, psychoanalysis was the primary approach used to classify and treat psychiatric illness. Psychoanalysis was used to describe underlying structural dysfunctions of the mind leading to psychiatric illness. As Freud stated, "Psychoanalysis is related to psychiatry approximately as histology is related to anatomy" (1917). For more than a generation, psychiatrists were trained in psychoanalysis, with the data derived from psychoanalytic techniques such as free association and dream interpretation used to make diagnoses and form treatment plans. Psychoanalysis was utilized with some success in treating psychiatric illness. However, most of the evidence attesting to the therapeutic efficacy of psychoanalysis was anecdotal and subjective (Grunbaum, 1984). There have been repeated failures to show that clinical outcomes achieved based on psychoanalysis differ from the outcomes achieved through supportive psychotherapy (Wallerstein, 1986). Psychoanalysis does differ from most other psychotherapeutic techniques in its attempt to offer insight into the basic psychodynamics of disease. The other approaches that have proven equal or better in the treatment of psychiatric illness include behavioral, cognitive, and supportive psychotherapy. These approaches limit their therapeutic goals to the elimination or improvement in unwanted or inappropriate behaviors.

Recently, applied psychoanalysis has been in serious decline as a method for diagnosing and treating psychiatric illness. In part this decline is due to socioeconomic considerations. There are many patients with psychiatric illness. Some surveys suggest that over a lifetime, almost 1/2 of Americans (46.4%) will meet criteria for a major psychiatric illness (America's Mental Health Survey, 2001). Psychoanalysis is costly and time consuming. Psychotherapy requires repetitive and long-term one-on-one therapy with a trained analyst. There are a limited number of psychoanalysts, and the training required for psychoanalysis is extensive. One trained analyst can treat only a limited number of patients. Based on cost–benefit requirements of the real world, psychoanalysis as a therapeutic technique has found it as primary application among the rich and famous.

Freud was without doubt a Monist. In fact psychoanalysis was developed in a predominately Cartesian era as a monist movement, an attempt to correlate mind processes with brain-based function. Through analysis of associative thought and dream content, Freud attempted to derive information about the structure of the brain from the study of aspects of the mind. Some of the problems intrinsic to applied psychoanalysis derive from attempts by Freud and other psychoanalysts to correlate their findings concerning psychic structure and dynamics with brain-based structures and functions. It was Freud's contention that dreams could be utilized as a window into the functioning of the brain. For the psychoanalyst working from a perspective based on the analysis of mind structure and function, the associative thought of dreams could be used to describe what could be known of the brain. For many, REMS (rapid eye movement sleep) = dreaming was the proof of this correlate. REM sleep is a process of the primitive brain stem. If dreaming and REM sleep are equivalent, through studying dreams the psychoanalyst could approach the basic drives of the primitive brain. The therapist could attempt to understand the dreams produced in the mythical "id." Psychoanalysts have been among those in the scientific and medical community who have tenaciously hung on to REMS = dreaming paradigm despite the mass of contrary evidence. For the psychoanalytic field, the loss of the REMS = dreaming correlate has been conceptually devastating for those who have proposed that psychoanalytic structures reflect actual brain structures (Fig. 22.2).

While the evidence is contradictory and limited for dreams being emanations of the id, there is even less evidence to support the neuroanatomical existence of other psychoanalytic mind-based structures. Psychoanalysis has fared poorly in the real world of clinical medicine. Even in the diagnosis and treatment of psychiatric illness, it has been difficult to integrate the successful use of psychoactive medications in the treatment of psychiatric illness with psychoanalysis. The medications or surgery utilized in treating medical illness are directed at the structural biological abnormality inducing the illness. It is rare for a psychiatric illness to have a known biological basis. The medications and treatments used to treat psychiatric illness treat the behaviors and complaints associated with the illness rather

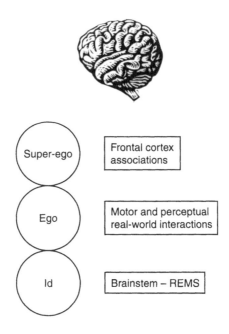

FIGURE 22.2 Super-ego, ego, id, and proposed neuroanatomical correlates.

than correcting known structural biological abnormalities. There are, however, underlying neurochemical abnormalities associated with a variety of psychiatric illnesses. Medications, such as antidepressants, alter the levels of active central nervous system (CNS) neurotransmitters such as serotonin, dopamine, and nor-epinephrine. Changes in active levels of these neurotransmitters can induce behavioral changes that can include the resolution and/or control of the psychiatric illness. Such neurochemical approaches to the treatment of psychiatric illness have proven to be far more successful, and less expensive therapeutic modalities than psychoanalysis. It is also far easier to train therapists to utilize these approaches. Today many in the fields of psychology and psychiatry are skeptical that psychoanalysis is a method that is useful in the diagnosis and treatment of psychiatric disease.

PSYCHOANALYSIS AND THE STRUCTURE OF MIND

Freud developed psychoanalysis as a Monist theory, an attempt to prove that the cognitive processes we call mind have a brain-based origin. No clear neurobiological correlates for psychoanalysis have been found. There is limited clinical proof that psychoanalysis results in positive therapeutic outcomes when used in the treatment of psychiatric disease. Despite extensive work in the field, there remains little evidence that psychoanalytic theories describe the psychodynamics that produce psychiatric illness. Psychoanalysis was developed as a diagnostic and therapeutic approach to treating psychiatric illness. A half-century of psychoanalytic focus on the treatment of psychiatric illness has led to the realization that there is little evidence to support this approach. Other psychodynamic and medical approaches to the treatment of these illnesses have proven both cheaper and more efficacious. The psychoanalytic era of psychiatry turned out to be one of long-term institutionalized therapy. The replacement of psychotherapeutic approaches with psychoactive medication and behavioral therapy has helped many psychiatric patients move out of the institutions and into functional lives. Applied generally throughout society, psychoanalysis proved to be an economic and social failure. Psychiatry as a field has rejected psychoanalytic approaches in order to embrace neurochemical and behavioral models of psychiatric disease. With the loss of such basic underpinnings to the field, it would not be surprising if psychoanalytic theories collapsed, and fell into disrepute.

The outcome has been quite different. During the same period in which applied psychoanalysis has been in serious decline as a psychotherapeutic method for treating illness, psychoanalysis has become one of the primary techniques utilized by individuals without diagnosed illness, in attempts their attempts to understand the structure and function of the mind. Psychoanalytic constructs of mind can be used to understand the content of dreams, the impulse and associative thought of creativity, the bi-directionality of cinema, and the impulsive power of art. In this era it is less likely that the trained psychotherapist will use psychoanalytic techniques. Psychoanalysis has moved to fields outside psychiatry that are also focused on attempting to understand aspects of higher cognitive functioning and mind. It is the well-read layman rather than the psychotherapist who is most likely to use psychoanalytic approaches and constructs in personal journeys of self-exploration.

CARTESIAN PSYCHOANALYSIS

Divorced from its connections to brain and psychiatry, psychoanalysis has become the most powerful theory that attempts to explain mind structure and function. Freudian and post-Freudian analytic perspectives seem to work far better off the couch, and outside the psychoanalyst's consulting room. The Cartesian psychoanalyst is more likely to become an analyst of mind rather than a therapist of disease. Freud and his accolades put much of their effort into extending the scientific method into the investigation of the mind. In that, they have been successful. Psychoanalysis has proven to be an excellent approach for attempting to understand mind-based cognitive processes. At various points in this book we have presented the evidence for the usefulness of Freudian and post-Freudian approaches in understanding cinema (Chapters 16 and 19), art (Chapter 20), creativity (Chapters 17 and 18), and the associative visual imagery and thought of dream (Chapters 12, 13, and 15). As we move in our investigation of dream science from basic neuroscience, through cognitive process, to higher levels of cognitive functioning, it becomes obvious that psychoanalytic perspectives work best when applied to higher cognitive functioning, those cognitive processes without clear brain-based correlates, those areas that we have been calling mind.

THE PSYCHOANALYTIC PERSPECTIVE OF MIND

Psychoanalysis provides us with at least a partial understanding of aspects of the mind. The techniques of psychoanalysis, free association, and dream interpretation, can be used to support structural patterns of mind. Structural constructs that have been developed in attempts to understand how cognitive aspects of mind function include the "super-ego, ego, id" structure of the mind developed by Freud (Fig. 22.3).

The portion of the mind that Freud called the id is an unconscious part of the mind embodying the instincts related to psychosexual gratification and operating without reference to the dictates of logic or external reality. The ego is wholly conscious, the instrument of learning, and adaptive relationship to the environment.

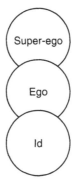

FIGURE 22.3 A model of psychoanalytic structure of the mind: super-ego, ego, and id.

The super-ego is part unconscious, operating as a monitor of conduct and functioning as a control, the nucleus of conscious providing the foundations for adult morality (Freud, 1923). This conceptual structure of the mind is utilized by many of the schools of psychoanalysis to psychoanalytically characterize the intrapersonal information obtained though dream analysis and free association.

Carl Jung extended this approach into the development of the concept of archetypes. Archetypes are socially and interpersonally shared symbols from mythology, often the same images experienced with emotional intensity in dreams and visions. Jung felt that these symbols describe shared aspects of the unconscious that connect an individual with the rest of mankind. An example is the abstract circle, utilized historically and religiously as an instrument of contemplation – the Mandela of Hindu and Buddhist religious art. Jung came to see the circle as an archetypical symbol of the self that includes the totality of both the conscious and unconscious psyche (1953). Such symbology is often present in dream content and imagery, and may reflect a normative description of how human beings see and understand the cognitive content of the mind.

Lacan and others have further developed post-Freudian constructs such as the "Other." Lacan's Other is the ego-ideal of our self-concept that we see reflected in a mirror. This is the understanding that we receive of ourselves through projection. This dramatically ambivalent ideal self-image is always outside us. Identification is a process in which the subject is displaced in an identification with the Other. Identification is a process with its own implicit ideology demanding sameness, necessitating similarity, and disallowing difference. These psychic constructs can be used to describe the relationship of the individual to imaginary and symbolic process (Kaplan, 1990).

No one has been able to demonstrate that these constructs of mind have actual correlates of brain structure and function. Psychoanalysis has proven to have limited usefulness in the field of medical or psychiatric disease, or when used to describe cognitive processes with clear brain-based correlates. However, these psychoanalytic constructs work fairly well when used to address cognitive processes without clear brain-based correlates – what we have been calling mind. Psychoanalytic constructs are commonly utilized in fields as diverse as film and art criticism, scientific methodology, dream content analysis, and studies of creative process and genius.

Psychoanalysis has a long history and a considerable literature that addresses the psychological basis and function of the process of associative thought. Freud's techniques of dream analysis have led to modern approaches using linguistic dream structure and content to describe the underlying narrative structure of thought (Foulkes, 1985; States, 1993). Freud's approach was surprisingly prescient in identifying the components of the dream state for which, even today, little brain-based evidence exists. There turns out to be describable brain/biological bases for both dream imagery and emotions. However, the biological basis for the narrative associative thought of dreams has yet to be described. Associative thought is a cognitive process that is somewhere beyond the limits of modern techniques of scientific analysis. Today the associative thoughts of dream are the components of dream that persist as a primary cognitive attribute of mind. These personally developed associative thoughts in dream frame the biologically based dream imagery, emotion, and memory.

Associative thought is also characteristic of creative process. Creativity is an associative process relying on spontaneous imagery and insights that may change and redefine the initial concept (Hunt, 1989). Freud suggested that we have the possibility of overcoming childhood traumas (bereavements, separations, aggressions, among others) through creative expression. Freud suggested that creations and works of art were imaginary satisfactions of unconscious wishes (1925). The part of creativity and art that are conceptually mind include the process involved in creating and appreciating a work of art, as well as the artistic product itself. Such an artistic creation of "meaning" from disparate unassociated memories is an associative process of imagery, emotion, and memory framed within thought, and can be considered one of the non-brain-based cognitive processes that we are calling mind.

Psychoanalysis has focused on the associative content of dreams. Frequent, intense dreaming, and creativity are associated with the same personality characteristics and psychiatric disorders. Both creativity and dreaming are global, gestalt self-expressive experiences diverging from the step-wise, perceptually defined consciousness in which we live our daily lives. Dreams, historically been a part of the paradigm for the creative process, are often used by individuals functioning successfully in creative roles. Psychoanalytic approaches to interpreting dream content move from the primarily visual dream to focus on associative memories and thoughts that are verbal and language based.

Psychoanalysts utilize these dream contents and associations in their attempts to understand the dynamics of mind-based process. As a result of this focus on dream interpretation, psychoanalytic constructs have become the primary approaches utilized in understanding dream content and associations. Psychoanalysis has proven more useful than any other approach in attempts to understand the intertwined cognitive processes of dreaming, creativity, and art.

In this book, we have focused on the creative process used in filmmaking. Freudian and post-Freudian analytic perspectives have become the primary approaches to films critique and analysis of the filmmaking process. Film can parody brain-based components of the dream state utilizing a visual storyline composed of associated images able to interact with the personal memories and emotions of the viewer. Film affects mind-based processes by integrating the viewer into its story, approaching the status of an almost complete cognitive experience through utilization of the same powerful cognitive processes that are utilized in dreaming. The Freudian apparatus theory, as well as, the post-Freudian structural perspectives of Lacan have proven useful as theoretical constructs available for describing and accounting for the interacting process of film and viewer.

Even in science, this psychoanalytic perspective has proven useful. Methodological problems explored in the psychoanalytic process have proven useful in examining problems inherent in the relationship between the researcher and the scientific experiment. If the results of a study are to be objectively replicable, results must not be obtained based on the desires and personal goals of the researcher or on the personal interactions between researcher and subject. Transference errors can occur even in this situation. Psychoanalysis has provided an approach useful in addressing the potential effects of self-reflection and personal knowledge of the clinician or researcher on the outcomes of scientific investigations (Frosh, 1998).

THE CARTESIAN PSYCHOANALYST

Psychoanalysis persists as a useful approach when applied to cognitive processes for which no clear biological basis can be described. It can be a questionable approach when applied to cognition with described biologic, brain-based correlates. Utilizing a Cartesian perspective of psychoanalysis, divorcing psychoanalysis from brain and applying these constructs to mind, has proven a useful approach useful for exploring the border of brain and mind. Areas of cognition in which psychoanalytic approaches work include creativity, associative thought and images, art, and feelings. These cognitive areas are the same as those for which neurological correlates are lacking. The areas in which psychoanalytic approaches apply are likely to be processes of mind. Areas of cognition where psychoanalytic approaches work less well include brain-based disease, motor activity, perception, communication, memory, and intelligence. These cognitive processes are likely to be processes of brain.

Psychoanalysis was developed as a scientific approach to the study of mind. It is only when freed of the need to identify biological bases for mind that psychoanalysis remains as a productive approach. The psychoanalyst approaches the border of brain from the perspective of mind. From this perspective, the brain appears global and diffuse with few identifiable specifics (Figs. 22.1 and 22.3). The cognitive rules affecting mind can be described and applied to mind-based processes, cognitive processes quite different to those specific processes defined by the process of sequential neurotransmission that apply on the other side of the Cartesian border between mind and brain.

There are advantages to adopting a Cartesian rather than a Monist perspective of mind and brain. Evidence is no longer ignored or denigrated and forced to fit theory. Scientific theories can reflect evidence. The study of REM sleep can be the study of an important electrophysiological state that affects learning and memory. The study of dream can be the study of cognitive reports of mentation from sleep involved in emotional and creative process. Neuroscientists can limit their theoretical constructs to what applies in the laboratory and in brain-based disease. The work and effort applied in the attempt to scientifically understand processes of mind through psychoanalysis can be applied to processes of mind rather than biologically based disease where it works less well.

Perhaps despite our anxiety and practical counter-intuition, it is better to assume the perspective that brain does not necessarily equal mind. Both neuroscience and psychoanalysis have a scientific and logical base. The neurosciences can be used to explain brain-based cognitive process. Psychoanalytic approaches can be used in attempts to understand processes of mind. Divorced from the need to explain the other, each starts to make logical sense, once again.

CHAPTER 23

Viewing the Border Between Brain and Mind: A Modified Cartesian Perspective

"What do you think would happen if the released prisoner went back to sit in his old seat in the cave? Would not his eyes become full of darkness, because he had come in suddenly out of the sunlight?"

"Certainly!"

"And if he again had to compete with the prisoners who were still shackled in giving opinions about the shadows, while he was still blinded and before his eyes got used to the dark – a process that would take some time – wouldn't he likely be set out to mockery? And wouldn't they say that he had only come back down to regain his eyesight, and that the ascent was not worth even attempting. And if anyone tried to release them [the prisoners] and lead them up, wouldn't they kill him if they could lay hands on him?"

"Certainly!"

(Plato's Cave – *The Fourth Stage – the Freed Prisoner's Return to the Cave* from *The Republic Book VII*, 516 e 3–517 a 6)

A border between brain and mind exists at the limits of current scientific knowledge in each of the fields of neurological and cognitive science. Scientific techniques can be utilized to explain the physical attributes and function of the primarily brain-based cognitive states involved in perceptual integration, computation, communication, emotions, and motor activity. For these cognitive states the argument can be made that brain = mind. However, even these primarily brain-based states have components that reach beyond brain-based activity to areas of cognitive functioning that are likely aspects of mind. There is motor activity that becomes dance. Communication that becomes song. Computation that utilizes gestalt-like integration in attempts to describe universal principles. Visual and auditory perceptions that are integrated and projected to become art. Emotions that become feelings.

The brain component is different for each of these cognitive states. The component of each state that can be considered as mind varies for each state as well as between individuals. This border between mind and brain is not static. The border for an individual can potentially change based on age, training, knowledge, life experience, mood, and even illness.

From the perspective of the primarily brain-based cognitive states, the border between mind and brain can appear impermeable, like a wall beyond which normal logic no longer seems to work. But like in an eclipse horizon, there is information that we can use in our waking functioning that leaks across this border. It seems likely that the information leaking from mind does not flow directly though the serial transmission-line neuro-processing that we utilize for interacting with the exterior world through motor activity, perceptions, and communication. The information that leaks across the border is more likely to arrive as part of the associative thought process characteristic of dreaming and utilized in creative process.

From the perspective of cognitive states that are primarily mind, it can be difficult, as well, to see across the mind/brain border to find components of each state that are brain based. But each mind-based state has a varyingly sized component that is clearly brain based. Feelings are tied to brain-based emotions. The visual and auditory arts require brain-based perceptual integration. To be successful in a creative role, such concrete measures as social and economic support and training in a field of expertise are required. Dreaming itself, that supposedly mind-based experience, utilizes brain-based visual systems and memories. Part of each cognitive state is brain. Part is mind. The components of these cognitive processes for which a neural basis has been demonstrated can be considered to be brain. The portions of these processes for which no neurological correlate has been demonstrated are considered to be mind. Each cognitive process is not fully either body or mind. There are components of each process that are brain-based and amenable to neurological technological analysis. There are components (mind) that are not. The proportions of each cognitive process considered to be either body or mind has considerable variation (Fig. 23.1).

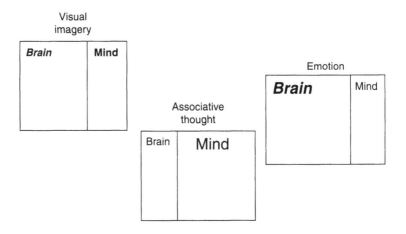

FIGURE 23.1 Modified Cartesian model of mind/body relationships for some of the cognitive processes involved in dreaming.

MODIFIED CARTESIAN MODELS

This modified Cartesian model as a bottom-up approach has certain advantages over concrete Monist or Cartesian approaches. This approach has a consistency with scientific evidence that neither the Monist nor the full Cartesian can boast. Within this model, the neuroscientist can restrict his or her work to scientific evidence for the function of the CNS based on repeatable experiment and medical pathology. There are no requirements that the same neuroscientist advance untestable theories reputed to explain aspects of mind. From the perspective of mind, this model is a top-down approach that can utilize current psychoanalytic models of mind structure and function. The modified Cartesian approach has the advantage of focusing studies of mind on those areas of each cognitive process that do not have a clear biological basis. The psychoanalyst can use that expertise in studying mind-based phenomena such as art, film, or personal creativity without the Monist requirement that the analysis demonstrate a neurological basis or a tie to psychiatric pathology.

We can use this modified Cartesian perspective to build a conceptual model of the mind/brain border cutting its path across the cognitive states (Table 23.1). Some states are predominately brain-based and can be created as artificial constructs

TABLE 23.1 A Modified Cartesian Perspective of the Proportional Representation of the Mind/Brain Border Across Selected Cognitive States

Brain

Cognitive state	Brain side	Mind side
Motor activity	**Brain**	*Mind*
Communication	**Brain**	*Mind*
Perception	**Brain**	*Mind*
Calculation	**Brain**	*Mind*
Focused waking thought	**Brain**	*Mind*
Memory	**Brain**	*Mind*
Intelligence	**Brain**	*Mind*
Emotion	**Brain**	*Mind*
Imagery	**Brain**	*Mind*
Problem solving	**Brain**	*Mind*
Dreaming	**Brain**	*Mind*
Associative thought	Brain	***Mind***
Creativity	Brain	***Mind***
Art	Brain	***Mind***

Mind

outside biological systems such as motor action, communication, and calculation. But even in these areas, biological systems such as human beings are drawn to aspects of mind, where motor activity becomes dance and communication becomes song. The border varies and changes like a fuzzy dotted line. That change will occur for the scientist as technology leads to an increased understanding of the extent of the brain basis for each of these states. That change occurs for each individual through the experience of life. The relative component of mind and brain, the looseness and/or concreteness of the border is likely to change, and to be different in differing individuals.

This is, of course, a theoretical model, and the proportions of brain versus mind ascribed to each cognitive state are approximations. The conceptual proportions of each state that are either mind or brain are based on this book's presentation of how mind and brain comprise each of the cognitive states. Tracing such variations in the extent of brain-based knowledge border across the different cognitive states allows us to attempt a visual model of the mind–brain border. None of the cognitive states are totally either brain or mind. There is a component of mind in the most concrete of physically based cognitive states (motor activity, perception, communication, and calculation) and a component of brain in the states that are most characteristic of mind (art and creativity).

We can attempt to fit this pattern into a visual paradigm. Mind and brain are two parts of a whole, and with attributes to Carl Jung, perhaps best visualized as a circle. The mind–brain border is unlike our visualization of the moon in its phases; there is never a full moon, a state that is totally mind or body, and there never a new moon state that has no component of either body or mind. The mind–brain border may be better described as components of a whole that are never complete.

One modeling approach that seems to work is derived from the round Yin/Yang symbol from Taoist philosophy. This ancient symbol utilized by ancient Chinese, Buddhist, Hindu, and early Christian religions and philosophies describes intertwined dualities in which each state includes a varying percentage and part of the other. The Yin/Yang symbol reflects a whole, an equilibrium between opposites in which neither quality is independent of the other. This symbol has added resonance in Western culture where the inverted form of the Yin/Yang with dark replaced with light has been adopted as the representative symbol for sleep medicine. Any diagonal line drawn across this symbol can be considered representative of a particular cognitive state where varying amounts of each state are either brain or mind (Fig. 23.2).

VISUALIZING THE MIND/BRAIN BORDER

From the brain-based scientific perspective, the border between mind and brain appears as a sharp delimiting line. The researcher using the tools of science and logic proceeds up to a point at which brain-based tools no longer work. Beyond that point is mind. From within a particular area of scientific focus, from such a brain-based perspective, the earth can appear to be quite flat. The brain-based scientist comes up against a sharp limit and reaches the edge of the concrete perceptual knowledge of the world available to scientific understanding. Beyond that point the scientist's tools either no longer work or give conflicting information. What information that is available is often qualitative and pictorial, sometimes giving

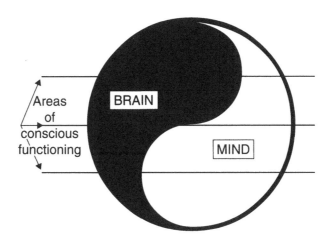

FIGURE 23.2   Yin/Yang symbol with diagonals representing areas of cognitive science.

the untestable sense of what may lie beyond. The philosopher Colin McGinn suggests that this inability to understand the mind using tools of science and logic may reside in the perceptual constructs that all of us utilize in representing external reality, "We should therefore be alert to the possibility that a problem that strikes us as deeply intractable, as utterly baffling, may arise from an area of cognitive closure in our ways of representing the world" (1991). He goes on to suggest that, "… our concepts of consciousness just are inherently constrained by our own forms of consciousness …" so that "… any theory that required us to transcend the finite genesis of our cognitive capacities would ipso facto be a theory we could not grasp … (1991)." It is McGinn's contention that mind states are not approachable with our logical and perceptual faculties. Mind may depend on the brain but mind cannot be observed using our brain-based perceptual introspective capabilities.

This is a problem we all seem to come to face when attempting to understand mind-based processes. The Monist moves on by ignoring contrary evidence and redefining the problem, all the while suggesting that the inexorable progress of science gives us the eventual capacity to solve any and all problems. However, it can be argued that the scientist who ignores evidence is no longer a scientist. The prisoner hiding in Plato's cave only wants to see the "positive" data supporting the reality of the existence where that individual has spent and focused their entire life. Plato viewed truth as the peeling back of layers of deception with the essence of truth being "deconcealment" (Heidegger, 1988). Behind each layer of explanation was not the final truth, not the grand and unifying theory of consciousness, but another layer with its own inherent shadows and deceptions. This is how the border of mind can appear to the brain-based scientist. Scientific and logical tools applied to one layer of explanation are likely not to apply in the next. Applied to the CNS, it seems likely that the next layer of truth will include broadcast electrophysiological rhythms and parallel neuro-processing. Scientific models and tools will be developed to address those processes and bring a bit more of cognition into the area of brain-based understanding. But beyond that layer will be another where the newest tools designed to scientifically explore that layer no longer seem to work. McGinn may be correct in suggesting that there will always be a component of cognitive functioning that we cannot scientifically or logically understand with the tools of physicality. There will always be mind.

Humans are not prone to accept limits. Humans willing to take the risks have left behind and ignored their concrete brain-based experience of the external world in order to explore the mind. Individuals have used religion, philosophy, meditation, psychoanalysis, and art as paths to be used in attempts to understand aspects of mind. Once there, they have sometimes tried to look back across the border at the brain. It is hard to see the brain from the perspective of mind, at least as hard as it was for us to see the mind from the perspective of brain. A focus on psychoanalytic theories of mind has been helpful in delineating processes of mind (areas of cognition where psychoanalytic approaches work) from processes of brain (areas of cognition where psychoanalytic approaches do not work). However, it has been as difficult for the psychoanalyst to demonstrate the brain-based basis his or her "art" as it has been for the neuroscientist to find mind-based correlates the electrophysiology and neural networks of his or her "science." Most depictions of the mind/brain border from the perspective of mind are quite different than those we obtain from the perspective of brain. The psychoanalyst studying the constructs of a patient's mind can perhaps obtain a diffuse understanding of brain-based medical disease. The artist can pull from creative aspects of mind ideas with which to structure a reflective simulacrum of perceptual reality. These are the aspects of brain derived from the perspective of mind. From the perspective of mind the border is diffuse and fuzzy, not the sharp and delineated border that we see from the perspective of brain. From the perspective of mind, our perceptual construct of the world appears quite different, a series of circles, archetypical symbols, or rules of perspective, one possible representation of an external reality that we do not fully apprehend.

DREAMING ACROSS THE BORDER

Dreaming is an example of a cognitive process that is both brain and mind-based. Dreaming extends across the border. We can scientifically understand the visual imagery, memories, and emotions involved in dreaming. We can use our brain scans and neuropathology studies to describe the brain sites at which these processes occur. We can describe the neural connections, electrophysiology, and the neurochemistry involved. But when we step beyond the border and attempt to use these techniques to understand the associative thought of dream. When we step further to try and understand that visceral shock of realization and connection that can come to us from mind, these brain-based techniques no longer work. There are creative insights and personal realizations available to us in our dreams and nightmares that we cannot scientifically or logically understand. Thoughts build new and unexpected associations. Here, brain scans are silent. Pathology studies are useless. Sequential neural connections may not even be required, over on the other side of dream across the border of brain and into mind.

CHAPTER 24

The Limits of Dream

I am sitting on a wooden bench. Embedded in the grain of the yellowed, varnished wood are tiny bright metallic colored, spot inlays. There are many different colors each representing a point of knowledge, a part of this book, each a tiny contributor to the understanding of dream. I rub the smooth surface of one inlay after another with my finger, satisfied that they fit together with a glittering beauty. The pattern formed by these glittering inserts is not geometric or mathematically correct. Instead they swarm in an arrangement defined and integrated with the used and gnarly grained wood of the bench. I scan further down the bench. Just beyond my reach there's a slightly larger inset, blue-gray in color. I feel that it is different, a hole in logic and understanding. I reach out to touch it. Somehow I can't. From that hole radiates a feeling of emptiness, loss and failure, of a nauseating spinning vertigo. I look around for help. My wife appears. I ask, "What is it? What have I missed?" A raven black horse comes out from around the corner and runs in front of the bench, full of energy, his tail up, his mane flying. I scan beyond him to a rolling farm of green pastures and trees searching for the limits of this place. "Is the fence down?" I wonder. "I need to get up and fix it before he escapes."

(Pagel, *Dream of this Book* (1/07))

It can be difficult to accept that there are limits in our ability to understand our way of existing and functioning in the world. Most of us focus on what we know rather than on what we do not understand. This book explores the contrary approach, studying what we do understand of an unusual interior mental experience that we call dreaming, and exploring the limits of current scientific knowledge of that cognitive state in order to identify what we do not understand. This is not a new approach. It is the same approach used by Descartes to develop the philosophical basis of modern scientific methodology (1637). It is one that has been used by philosophers and scientists since the dawn of civilization. Most of us dream. Most of us wonder what our dreams may mean.

There are portions of dream that we can scientifically understand. We understand the neural organization of visual perception better than most other cognitive process. We use this system to form the imagery of our dreams. These dream images come out of our visual memories of past experience, the same memories that are accessible during waking. Neuroscientists are beginning to understand the outlines of the neural basis of at least some of these memory systems. We know that emotions have biologic and evolutionary functions. We are starting to understand the neural basis of emotions and how dreaming may function in emotional processing. Dreaming is generally visual and incorporates experiential memories. Dreams are often emotional, involved in the processing of waking emotional experience. It is on these brain-based systems that our dreams are built.

Is this what a dream is? Can dreaming can be fully characterized by these biological components? We can create artificial dreams that affect our perceptions, memories, and emotions. The filmmaker takes a personal story, and presents that story visually, attempting to maintain the attention of the viewer. Visual images presented on film, like the visual images of dream, are outside the viewer's conscious control. The powerful effects that cinema can have on a viewer derive in

FIGURE 24.1 Dream image of bench.

part from its ability to access the same visual, emotion, and memory systems that we use in dreaming. This parody of dream can affect the emotions of the viewer, going so far as to access a viewer's personal memories that resonate with the story and images of film. The filmmaker can create a shared dream, consciously using genre, archetypical symbols, and mythological allusions in attempts to make contact with the viewer. But often the response to a film is different than the filmmaker ever would have ever expected. Filtered through personal associations, memories, and experience the film becomes for the viewer an experience that resembles a dream (Fig. 24.1).

BRAIN-BASED CORRELATES OF MIND

There are, however, cognitive processes that cannot be explained by current scientific technology. Cognitive scientists have used current knowledge of neuroanatomy, neuroscanning, neurochemistry, and electrophysiology to develop testable scientific theories explaining brain-based cognitive processes. That approach has worked poorly when applied to the cognitive states without clear brain-based origin. The most notable example of such a theoretical construct has been the correlation of rapid eye movement (REM) sleep with dreaming. For the last 50 years, REMS = dreaming has been generally accepted as scientific evidence for a brain-based correlate of mind. In order to support this theory, overwhelming contrary evidence has been ignored, along with the objections of many, if not most, of the major researchers involved in the scientific study of dreaming.

Other potential brain-based correlates of mind exist. Most of these correlates depend on using limited applied definitions for "mind." There are brain-based correlates of emotion and of imagery. There are even components of dreaming that are brain based. If emotion, imagery, and dreaming are defined as mind, then the evidence exists that brain equals mind. This approach works much as blinders work for the speeding race horse, curtailing alternative perspectives that interfere with the concrete task at hand (Fig. 24.2). However, with blinders off, there are few cognitive scientists who would argue that the cognitive operation and function of the central nervous system (CNS) can be fully explained within current scientific technological paradigms of brain structure. Neuroscientists tend to emphasize what we do know about brain functioning and de-emphasize the huge amount of CNS functional operation that we do not understand. Neurotransmission theory explains the neural connections and brain structure required for brain-based cognitive functions such as motor activity, perception, communication, and calculation. Classic neuroanatomy based on sequential transmission theory has been far less successful in demonstrating brain-based correlates for thought, particularly associative thought. Practitioners and theorists of sequential neuroscience tend to ignore how it might be applied to other mind-based cognitive process including the integration of associative visual imagery, creativity, art, and feelings (Young, 1987). No one has been able to correlate these cognitive processes with specific brain sites, switches, or neurochemical triggers. These cognitive processes including thought, particularly associative thought and images, creativity, art, and feelings are examples of what in this book we have been calling "mind."

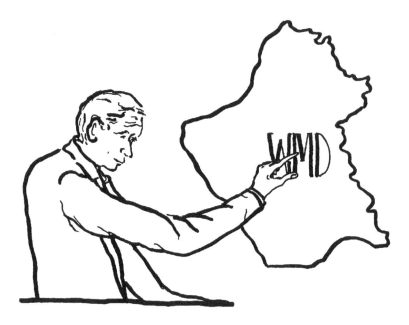

FIGURE 24.2 One approach to "scientific" theories of cognition: redefinition with blinders.

MIND-BASED COGNITIVE PROCESSES

Mind-based processes have considerable importance. These processes are beyond the capacity of artificial intelligence systems. Without these processes human beings can be modeled as biological systems operating as automatons. These are the "higher" level functions utilized by the conscious individual to maintain functional capacity in a derivative perceptual representation of the external world. These are the systems that integrate and associate disparate biological brain systems into the complex phenomenon that we call consciousness.

Parallel neuro-processing theories suggest that the CNS may function in a non-sequential manner. If as seems likely, such neural function does characterize CNS operations, the CNS is even more complex that initially hypothesized. Complexity theory suggests the possibility that consciousness naturally develops in any system that is pushed to a certain point of complexity. Complexity, to some extent a Newtonian theory, suggests that if we can diagram all sites and potential interactions of neurons, we can create artificially consciousness systems that parody the function of the human CNS. If this is true, based on the extreme cosmological complexity of the universe, there are likely many conscious systems that have obtained the level of complexity attainable by the human CNS.

The complexity = conscious theory suffers somewhat when applied to the electrical power grid attached to this personal computer. This system when integrated with others has an extreme level of complexity. Yet the pattern of electrical cost, control, and potential for miscalculation involved in the functioning of our local and global electrical networks seems very human. Although this is a sequential system, I cannot help but think that an independently conscious system would function differently, perhaps even better than the infrastructure that we have invented and maintained.

There may be other brain systems have the capacity to support global CNS functions. Scientists are just beginning to explore the background physiologic electrical rhythms of the EEG. The activity of this system is most apparent during sleep. These rhythms are associated with dreaming, and individuals trained in meditation can control and alter the presentation of these frequencies. This global system reflects the coordinated electrical discharge of large populations of neurons and other electrically sensitive neural cells. There is increasing evidence for the functioning of this system. Such physiological electrical rhythms are known to affect the tendency of neurons to fire. Abnormalities of this system are associated with disease states including attention deficit/hyperactivity disorder and pediatric obstructive sleep apnea. This system has the capacity to affect cellular functioning providing both information and energy to individual neurons. Electroneurochemical receptors exist on the cellular membrane with the potential to affect the expression of DNA, one of the most electrically sensitive of proteins. Such an extracellular electrical system has the potential to assess the cellular memory stored in DNA. Such a system could be utilized in global mind-based cognitive states.

These theoretical constructs expand the potential for explaining components of CNS function that cannot currently be explained by sequential neurotransmission. If as seems likely, DNA-based cellular memory is actually integrated with CNS neuro-processing, the complexity of the CNS increases exponentially beyond the complexity of current models. That complexity is already extreme, and modeling such a system is beyond the capacity of current artificial systems. Simple top-down models have been of some use in the development of neuroscience, but the CNS as we understand it today is not in any way simple. When the evidence no longer supports such top-down models, when such models become misleading and distracting, they should be discarded. There are nuggets of truth embedded in the structures of activation–synthesis, AIM, and focused attention. Those insights should be kept as evidence suggests that they still apply.

Without the support of brain-based correlates, many expected that psychoanalysis would deconstruct in this modern era of cognitive neuroscience. Instead psychoanalytic theories have persisted, showing continued value as techniques of scientific analysis when applied outside the psychotherapeutic focus. Psychoanalytic models still work better than any other current model when applied to associative thought, personal growth, art, and film studies. Freud's highly developed legacy of psychoanalytic methodology and theory still has much to offer when applied to cognitive aspects of mind.

Thought can be approached as the process utilized to frame imagery within a narrative storyline. Thought denotes personal presence in the cognitive process whether it be dream, art, or creativity. As in literature, thought provides authorial direction, a framework for what would otherwise be the limited, two-dimensional images characteristic of waking imagery and dream (Scarry, 1995). Thought denotes the personal presence of the individual in the process of imaginative reconstruction. Creative associative thought can frame imagery in a way that the vivacity, solidity, and persistence of the perceptual world can be created and even surpassed for the dreamer, the artist, the creator, the reader, or the viewer of art. During the dream state, entire, seemingly real, perceptual universes can be recreated.

In dream thought provides a mind-based framework for brain-based imagery, memory, and emotion. The process of thought denotes the personal cognitive presence of the individual having the dream. Associative thought integrates imagery, emotion, and memory into a story that can be remembered and reported in waking. It is thought that has the capacity to create "meaning" from disparate unassociated memories. This process of associative thought in dreaming is what makes dreams "significant," and powerful enough to disrupt sleep and affect waking attitudes and behaviors. Of all the cognitive processes involved in mentation occurring during sleep, the state of consciousness called dreaming, it is thought that can most clearly be considered to be mind.

THE LIMITS OF KNOWLEDGE: BORDERS

The requirement that mind equals brain is a basis for current top-down psychoanalytic and neuroscientific theories. Yet, it is difficult to visualize the mind from the perspective of brain. And it is even more difficult to visualize the brain from the perspective of mind. From the perspective of mind, the brain basis for psychoanalytic psychiatry is not apparent. From the perspective of brain, mind-based correlates for electrophysiology and neural networks have turned out to be illusions. Adherents to the major theoretical perspectives requiring that mind = brain have tended to reject the evidence supporting both of these truths. It is a fact that REMS does not equal dreaming. This is no longer a question of scientific or logical evidence. For the adherents of the theories of cognitive neuroscience based on activation–synthesis, and for the practitioners of psychoanalytic psychiatry, it is a question of belief.

It is difficult to accept that there may be a border that limits our ability to understand our way of existing and functioning in the world. Yet, there is still much that we do not understand about CNS functioning. It would be foolish not to consider the possibility that there are aspects of human CNS functioning that we, as humans, may never be unable to understand. The border between mind and brain may be an eclipse horizon that cannot be crossed from the perspective of either mind or brain.

Colin McGinn suggests that human beings may not have the capacity to understand mind-based cognitive processes. "… we are cut off by our very cognitive constitution from achieving a conception of that natural property of the brain (or of consciousness) that accounts for the psychophysical link" (McGinn, 1989). Once we choose to approach mind from the perspective of brain-based science we are locked into the perspective obtainable from that side of the border. McGinn postulates that the techniques of logical and scientific analysis based on the derivative perceptual processes that we use to understand the world outside us no longer work when applied to mind. "Everything physical has a purely physical explanation. So the property of consciousness is cognitively closed with respect to the introduction of concepts by means of inference to the best explanation of perceptual data about the brain" (McGinn, 1989). Conceptually, this understanding of mind-based process is built on as much logic and evidence as current brain-based theories purporting to explain the basis of mind.

At the limit of dream there is an event horizon. This border where scientific evidence can no longer support our brain-based theories of cognitive function is the border between brain and mind. This border exists for all cognitive states, even for those that are primarily brain based. Dreams leak through this border, built on brain-based neurotransmission systems of experiential perceptual memory, imagery, and emotion. The dream comes as a cascade of thoughts and images leading to further associations of thought and image that integrate into the storyline of the dream. These dream thoughts have meaning for the dreamer, often a complexity of association and integration that on analysis can be more complex than waking thought. These insights of dream can have the intensity of dream. Experiences of waking life only rarely have the intensity or power, the discombobulating sense of reality that one can experience in a dream.

When I experienced the totality of the solar eclipse in Hawaii, I was struck by the advance of the eclipse horizon as it charged across the earth, hitting me with the psychological force of a locomotive. From one perspective it was nothing more than the change of light to dark. From another, it was an intensely visceral experience that left me exhausted. Such images in metaphor are the stuff of dreams, the glittering blue-gray insert into a sitting bench of dream reflects an aspect of my mind that leaves me on waking with physical nausea and disorientation. This dream had the capacity to affect my goals and plans and even the structure of this thesis. My personal response to this dream was to search waking associations and memories, and indulge myself, lapsing into personal psychoanalysis. I checked both the fence at my ranch and reviewed the structure of the book. I discussed the dream with my wife and friends who pointed out obvious connections unapparent to me as the dreamer. I see a similar bench in a book and start to draw up plans. The bench will be in this book. The bench will be in my new house.

Such dreams tunnel across the eclipse horizon that marks the border of mind, in defiance of both physics and neuroanatomy. Such dreams seem to come from outside the sequential concrete processing of my neurotransmitting brain. In waking, the possible associations are endless, and the potential levels of analysis multiple and profound. In my future, images from this dream will resonate and have their own reflections. Such dreams are remarkable personal experiences, intuitive insights, benchmarks, even art – stories that come from beyond the biological limits of dream and lead to creative insight. Such dreams are aspects of mind that crash across a seemingly impenetrable border into the waking world. I sit here among my books, working at my computer, and look toward that limiting border lying beyond intellect and science, wondering just what else is there.

REFERENCES

CHAPTER 1

Descartes, R. (1641). Objections against the meditations and replies. In *Great Books of the Western World: Bacon, Descartes and Spinoza (1993)*, Editor-in-Chief M. J. Adler. Encyclopaedia Britannica, Inc., Chicago, IL, pp. 17, 21.

Descartes, R. (1980). *Discourse on Method (1637) and Meditations on First Philosophy (1641)*, trans. D. A. Cress. Hackett Publishing Company, Indianapolis, IN.

Flanagan, O. (1992). *Consciousness Reconsidered*. The MIT Press, Cambridge, MA, p. 116.

Green, B. (2003). *The Elegant Universe: Superstrings, Hidden Dimensions, and the Quest for Ultimate Theory*. W. W. Norton & Company, NYC, NY, p. 344.

Hawking, S. (1993). *Black Holes and Baby Universes and Other Essays*. Bantam Books, New York, pp. 19, 108, 125.

Heisenberg, W. (1952/1979). *Philosophical Problems in Nuclear Science*, trans. F. Hays. Oxbow Press, Woodbridge, CT, p. 104.

Luminet, J. (1992). *Black Holes*. Cambridge University Press, Cambridge.

McCarley, R. and Hobson, J. (1975). Neuronal excitability modulation over the sleep cycle: A structural and mathematical model. *Science* 189: 58–60.

McGinn, C. (1989). Can we solve the mind–body problem? *Mind* 98: 349–366.

McGinn, C. (1991). *The Problem of Consciousness*. Blackwell, Oxford, p. viii.

Nagel, T. (1974). *Mortal Questions*. Cambridge University Press, Cambridge, p. 47.

Neilsen, T. (2003). A review of mentation in REM and NREM sleep: "Covert" REM sleep as a possible reconciliation of two opposing models. In *Sleep and Dreaming: Scientific Advances and Reconsiderations*, eds. E. Pace-Schott, M. Solms, M. Blagrove and S. Harnad, Cambridge University Press, Cambridge, UK, pp. 59–74.

Pagel, J. (2003). Dreaming is not a non-conscious electrophysiologic state. In *Sleep and Dreaming: Scientific Advances and Reconsiderations*, eds. E. Pace-Schott, M. Solms, M. Blagrove and S. Harnad, Cambridge University press, Cambridge, UK, pp. 196–200.

Pinker, S. (1997). *How the Mind Works*. W. W. Norton, New York, p. 77.

SECTION 1

Descartes, R. (1641). Objections against the meditations and replies. In *Great Books of the Western World: Bacon, Descartes and Spinoza (1993)*, Editor-in-Chief M. J. Adler. Encyclopaedia Britannica, Inc., Chicago, IL.

Mahoney, P. (1987). *Freud as a Writer*. Yale University Press, New Haven and London.

CHAPTER 2

Blumenfield, D. and Blumenfield, J. B. (1978). Can I know that I am not dreaming? In *Descartes: Critical and Interpretive Essays*, ed. M. Hooker. The Johns Hopkins University Press, Baltimore, pp. 235–255.

Churchland, P. (1986). *Neurophilosophy: Toward a Unified Science of the Mind/Brain*. The MIT Press, Cambridge, MA, p. 272.

Damasio, A. (1994). *Descartes' Error: Emotion, Reason and the Human Brain*. Avon Books Inc., New York.

Descartes (1637). Disourse on method, trans. D. A. Cress (1980). Hackett Publishing Company, Indianapolis, IN.

Descartes, R. (1641). Objections against the meditations and replies. In *Great Books of the Western World: Bacon, Descartes and Spinoza (1993)*, Editor-in-Chief M. J. Adler. Encyclopaedia Britannica, Inc., Chicago, IL, pp. 68, 78, 100.

Descartes, R. (1980). *Discourse on Method (1637) and Meditations on First Philosophy (1641)*, trans. D. A. Cress. Hackett Publishing Company, Indianapolis, IN.

Freud, S. (1923). *The Ego and the Id*, trans. 1927, London; reprinted in Complete Psychological Works, Vol. XIX (1961).

Hobson, J. (1994). *The Chemistry of Conscious States*. Little, Brown and Company, Boston, MA, p. 205.

Jouvet, M. and Michael, F. (1959). Correlation electromyographiques du sommeil chez le chat decortique et mesencephalique chronique. Comptes Rendus des Seances de La Societe de biologie et de Ses Filiates 153: 895–899.

Kim, J. (1998). *Philosophy of Mind*. Westview Press, Boulder, Colorado, pp. 47, 52.

McGinn, C. (1991). *The Problem of Consciousness*. Blackwell, Oxford.

McGinn, C. (2002). *The Making of a Philosopher: My Journey through Twentieth-Century Philosophy*. Harper Collins, New York.

Place, U. (2004). *Identifying the Mind*, eds. G. Graham and E. Valentine. Oxford University Press, Oxford.

Plato, Plato's Cave – the First Stage, from The Republic, Book VII, 514a–517a, taken from Platonis Oprea, recogn. Ioannes Burnet, Oxonii: clarendon, 2nd Edition (1905–1910) Vol. 4 (English trans.) M. Heidegger included in The Essence of Truth (2002) Continuum, London.

Schouls, P. (2000). *Descartes and the Possibility of Science*. Cornell University Press, Ithaca, NY, p. 152.

The Gospel of Thomas: The Nag Hammadi Text [29] approximately 100 AD.

Vrooman, J. R. (1970). *Rene Descartes: A Biography*. G. P. Putnam's Sons, New York.

Webster's New Universal Unabridged Dictionary (1996). Barnes and Nobel Books, New York.

CHAPTER 3

Coleridge, S. (1912/1979). *Coleridge: Poetical Works*, ed. E. Coleridge. Oxford University Press, Oxford.

Dead Sea Scrolls (180 BC), J. Sira, "Qumran Dead Sea Scrolls, Apochypha, sirach 34.3." In *The New Oxford Annotated Bible with the Apocrypha* (1973), ed. H. May and B. Metzger. Oxford University Press, New York, p. 172.

de la Baroa, P. (1670). Life is a Dream, Act II, p. 119.

Foulkes, D. (1985). *Dreaming: A Cognitive–Psychological Analysis*. Lawrence Erlbaum Associates, Hillsdale, NJ.

Francis, A. (Chair – Task Force on DSM-IV) (1994). *DSM-IV – Diagnostic and Statistical Manual of Mental Disorders*, 4th Edition. American Psychiatric Association, Washington, D.C.

Freud, S. (1953). *The Interpretation of Dreams*, J. Strachey (ed.) The Standard Editions of the Complete Psychological Works of Sigmund Freud, Vol. IV and V. (1907). Hogarth Press, London, England.

Giambra, L., Jung, R. and Grodsky, A. (1966). Age changes in dream recall in adulthood. *Dreaming* 6(1): 17–32.

Globus, G. (1991). Dream content: Randm or Meaningful? *Dreaming* 1(1): 27–40.

Hadfield, J. A. (1954). *Dreams and Nightmares*. Baltimore, Penguin books, pp. 146–147.

Hartmann, E. (1991). *Boundaries of the Mind*. Basic Books, New York.

Hartmann, E. (1998). *Dreams and Nightmares – The New Theory on the Origin and Meaning of Dreams*. Plenum Trade, New York and London, p. 90.

Hill, C., Diemer, R., Hess, S., Hillyer, A. and Seeman, R. (1993). Are the effects of dream interpretation on session quality, insight, and emotions due to the dream itself, to projection, or to the interpretation process? *Dreaming* 3(4): 269–280.

Hobson, J. A. (1988). *The Dreaming Brain*. Basic Books, New York.

Hunt, H. (1991). Dreams as literature/science of dreams: An Essay. *Dreaming* 1(3): 235–242.

Jung, C. (1965). *Memories, Dreams, Reflections*. Vintage, New York.

Krippner, S., Gabel, S., Green, J. and Rubien, R. (1994). Community applications of an experimental approach to teaching dreamwork. *Dreaming* 4(4): 215–222.

Levin, R. (1994). Sleep and dreaming: Characteristics of frequent nightmare subjects in a university population. *Dreaming* 4(2): 127–138.

Mahowald, M., Woods, S. and Shenck, C. (1998). Sleeping dreams, waking hallucinations and the central nervous system. 8(2): 89–102.

Moffittt, A., Kramer, M. and Hoffman, R. (eds.) (1993). *The Functions of Dreaming*. State University of New York Press, Albany, NY.

Pagel, J. F. (1999). A dream can be gazpacho – searching for a definition of dream. *Dreamtime* 16(1&2): 6–28.

Pagel, J. F. (Chair), Blagrove, M., Levin, R., States, B., Stickgold, B. and White, S. (2001). Defining dreaming – a paradigm for comparing disciplinary specific definitions of dream. *Dreaming* 11(4): 195–202.

Pagel, J. and Vann, B. (1992). The effects of dreaming on awake behavior. *Dreaming* 2(4): 229–238.

Pagel, J. F., Crow, D. and Sayles, J. (2003). Filmed dreams: cinematographic and story line characteristics of the cinematic dreamscapes of John Sayles. *Dreaming* 13(1): 43–48.

Random House Dictionary of the English Language – 2nd Edition Unabridged (1987), Editor-in-Chief J. Stein. Random House Inc., New York.

States, B. (1992). The meaning of dreams. *Dreaming* 2(4): 249–262.

Stefanakis, H. (1995). Speaking of dreams: A social constructions account of dream sharing. *Dreaming* 5(2): 95–104.

Tedlock, B. (1991). The new anthropology of dreaming. *Dreaming* 1(2): 161–178.

Visson, L. (1992). Moments of truth: Dreams in Russian literature. *Dreaming* 2(3): 181–190.

Webb, W. (1992). A dream is a poem: A metaphorical analysis. *Dreaming* 2(3): 191–202.

CHAPTER 4

Armitage, R. et al. (1994). The effects of nefazodone on sleep architecture in depression. *Neuropsychopharmachology* 10(2): 123–127.

Benson, D. F. (1994). *The Neurology of Thinking*. Oxford University Press, New York, pp. 113, 117.

Dewan, E. M. (1970). The programming (P) hypothesis for REM sleep. *International Psychiatry Clinics* 7(2): 295–307.

Domhoff, W. (2003). *The Scientific Study of Dreams: Neural Networks, Cognitive Development and Content Analysis*. American Psychological Association, Washington, D.C.

Foulkes, D. (1985). *Dreaming: A Cognitive–Psychological Analysis*. Lawrence Erlbaum Associates, Hillsdale, NJ.

Hall, C. and Van de Castle, R. (1966). *The Content Analysis of Dreams*. Appleton-Century-Crofts, New York.

Hobson, J., Pace-Schott, E. and Stickgold, R. (2003). *Dreaming and the Brain: Toward a Cognitive Neuroscience of Conscious States in Sleep and Dreaming: Scientific Advances and Reconsiderations*, Cambridge University Press, Cambridge, UK. eds. E. Pace-Schott, M. Solms, M. Blagrove and S. Harnad. pp. 1–50.

Hobson, J. A. (1994). *The Chemistry of Conscious States – How the Brain Changes Its Mind*. Little, Brown and Company, Boston, MA.

Morrison, A. and Reiner, P. (1985). A dissection of paradoxical sleep. In *Brain Mechanisms of Sleep*, eds. D. Mcginty, R. Drucker-Colin, A. Morrison and P. Parmeggiani. Raven Press, New York.

Myers, P. and Pagel, J. F. (2001). The effects of daytime sleepiness and sleep onset REMS period (SORP) on reported dream recall. *Sleep* 24: A183.

Neilsen, T. (2003). A review of mentation in REM and NREM sleep: "Covert" REM sleep as a possible reconciliation of two opposing models. In *Sleep and Dreaming: Scientific Advances and Reconsiderations*, Cambridge University Press, Cambridge, UK. eds. E. Pace-Schott, M. Solms, M. Blagrove and S. Harnad. pp. 59–74.

Nofzinger, E., Mintun, M., Wiseman, M., Kupfer, D. and Moore, R. (1997). Forebrain activation in REM sleep: An FDG PET study. *Brain Research* 770: 192–201.

Pace-Schott, E. (2003). Postscript recent finding on the neurobiology of sleep and dreaming. In *Sleep and Dreaming: Scientific Advances and Reconsiderations*, eds. E. Pace-Schott, M. Solms, M. Blagtove and S. Harnad. pp. 335–350.

Pagel, J. F. and Vann, B. (1995). Polysomnographic correlates of reported dreaming: Negative correlation of RDI with dreaming effects on waking activity. *APSS Abstracts*. p. 51.

Pagel, J. F., Pegram, V. et al. (1995). REM sleep and intelligence in mice. *Behavioral Biology* (9): 383–388.

Roffwarg, H., Muzio, J. and Dement, W. (1966). Ontogenetic development of the human sleep–dream cycle. *Science* 152: 604–619.

Shankara (788–820 A. D.) Viveka Chudamani (Vedic Scriptures). In *The Pinnacle of Indian Thought* (1967) ed. E. Wood. Theosophical Publishing House, Wheaton, Illinois.

Smith, C. (1985). Sleep states and learning: A review of the animal literature. *Neuroscience and Behavioral Reviews* 9: 157–168.

Smith, C. (1995). Sleep states and memory processes. *Behavioral Brain Research* 69(1–2): 137–145.

Solms, M. (1997). *The Neurophysiology of Dreams: A Clinico-anatomical Study*. Lawrence Erlbaum, Mahwah, NJ.

Solms, M. and Turnbull, O. (2002). *The Brain and the Inner World: An Introduction to the Neuroscience of Subjective Experience*. Other Press, New York, pp. 141–143, 204.

Vertes, R. (1986). A life sustaining function for REM sleep: A theory. *Neuroscience and Biobehavioral Reviews* 10: 371–376.

Wasserman, M., Pressman, M., Pollak, C., Spielman, A., DeRosario, and Weitzman, E. (1982). Nocturnal penile tumescence: Is it really a REM phenomenon? *Sleep Research* 11: 44.

Winson, J. (1972). Interspecies differences in the occurrences of theta. *Behavioral Biology* 7: 479–487.

CHAPTER 5

Coupland, N. J., Bell, C. J. and Potokar, J. P. (1996). Serotonin reuptake inhibitor withdrawal. *Journal of Clinical Psychopharmacology* 16(5): 356–362.

Dimsdale, J. and Newton, R. (1991). Cognitive effects of beta-blockers. *Journal of Psychosomatic Research* 36(3): 229–236.

Gursky, J. and Krahn, L. (2000). The effects of antidepressants on sleep: A review. *Harvard Review of Psychiatry* 8(6): 298–306.

Hobson, J., Pace-Schott, E. and Stickgold, R. (2003). *Dreaming and the Brain: Toward a Cognitive Neuroscience of Conscious States in Sleep and Dreaming: Scientific Advances and Reconsiderations*, Cambridge University Press, Cambridge, UK. eds. E. Pace-Schott, M. Solms, M. Blagrove and S. Harnad. pp. 1–50.

Hobson, J. A. (1999). *Dreaming as Delerium*. The MIT Press, Cambridge, MA.

Hobson, J. A. and Steriade, M. (1986). The neuronal basis of behavioral state control: Internal regulatory systems of the brain. In *Handbook of Psychology*, Vol. IV(14), Mountcastle V, Series Edition, ed. F. Bloom. American Physiological Society, Washington, D.C., pp. 701–823.

Jaffe, S. E. (2000). Sleep and infectious disease. In *Principles and Practice of Sleep Medicine*, Vol. 3, eds. M. Kryger, T. Roth and W. Dement. W. B. Saunders Co., Philadelphia, PA, pp. 1093–1102.

Jones, B. (1998). The neural basis of consciousness across the sleep-waking cycle. In *Consciousness: At the Frontiers of Neuroscience, Advances in Neurology*, Vol. 77, eds. H. Jasper, L. Descarries, V. Castellucci and S. Rossignol. Lippincott-Raven, New York.

Kandle, E. R. (2000). The brain and behavior. In *Principles of Neural Science*, 4th Edition, eds. E. R. Kandel, J. H. Schwartz and T. M. Jessell. McGraw Hill, New York, pp. 5–18.

Krissel, J. et al. (1994). Thiopentone, thiopentone/ketamine, and ketamine for induction of anesthesia in caesarean section. *European Journal of Anaesthesiology* 11: 115–122.

Krueger, J. M. and Fang, J. (2000). Host defense. In *Principles and Practice of Sleep Medicine*, Vol. 3, eds. M. Kryger, T. Roth and W. Dement. W. B. Saunders Co., Philadelphia, PA, pp. 255–265.

Krueger, J. M., Kubillis, S., Shoham, S. et al. (1986). Enhancement of slow wave sleep by endotoxin and libid A. *American Journal of Physiology* 251: R591–R597.

Lepkifkier, E. et al. (1995). Nightmares related to fluoxetine treatment. *Clinical Neuropharmacology* 18(1): 90–94.

Mallick, B. N. and Saxena, K. S. (2001). *Interactions between cholinergic and GABAnergic neurotransmitters in and around the locus coeruleus for the induction and maintainence of rapid eye movement sleep. Neuroscience* 104: 467–485.

Marsh, S. et al. (1992). Dreaming and anesthesia: total i.v. anaesthesia with propofol versus balanced volatile anaesthesia with enflurane. *European Journal of Anaesthesiology* 9: 331–333.

McCarley, R. and Hobson, J. (1975). Neuronal excitability modulation over the sleep cycle: A structural and mathematical model. *Science* 189: 58–60.

Mitler, N. M. (2000). Nonselective and selective benzodiazepine receptor agonists: Where are we today? *Sleep* 23(Suppl 1): S39–S47.

Oxorn, D. et al. (1997). The effects of midazolam on propofol-induced anesthesia: Propofol dose requirements, mood profiles, and perioperative dreams. *Anesthesia and Analgesia* 85: 553–559.

Pace-Schott, E. et al. (1999). Effects of serotonin reuptake inhibitors (SSRI) on dreaming in normal subjects. *Sleep* 22(Suppl 1): H278D.

Pace-Schott, E. F. (2003). Postscript: Recent findings on the neurobiology of sleep and dreaming. In *Sleep and Dreaming: Scientific Advances and Reconsiderations*, eds. E. F. Pace-Schott, M. Solms, M. Blagrove and S. Harnad. Cambridge University Press, Cambridge, pp. 335–350.

Pagel, J. F. (2006a). Medications that can cause sleepiness. In *Sleep a Comprehensive Handbook*, ed. T. Lee-Chiong. J. Wiley and Sons, Hoboken, NJ, pp. 175–182.

Pagel, J. F. (2006b). The neuropharmacology of nightmares. In *Sleep and Sleep Disorders: A Neuropsychopharmacologic Approach*, eds. S. R. Pandi-Perumal, D. P. Cardinali and M. Lander. Landes Bioscience, Georgetown, TX, pp. 225–240.

Pagel, J. F. and Helfer, P. (2003). Drug induced nightmares – An etiology based review. *Human Psychopharmacology Clinical Experience* 18: 59–67.

Perry, E. and Perry, R. (1995). Acetylcholine and hallucinations: disease-related compared to drug-induced alterations in human consciousness. *Brain and Cognition* 28: 240–258.

Schwartz, J. H. (2000). Neurotransmitters. In *Principles of Neural Science*, 4th Edition, eds. E. R. Kandel, J. H. Schwartz and T. M. Jessell. McGraw Hill, New York, pp. 280–297.

Stacy, M. (1999). Managing late complications of Parkinson's disease. *Medical Clinics of North America* 83(2): 469–481.

Steriade, M. (2000). Brain electrical activity and sensory processing during waking and sleeping states. In *Principles and Practice of Sleep Medicine*, 3rd Edition, eds. M. H. Kryger, T. Roth and W. C. Dement. W. B. Saunders Co., Philadelphia, PA, pp. 93–111.

Thompson, D. and Pierce, D. (1999). Drug induced nightmares. *The Annals of Pharmacotherapy* 33: 93–96.

======== ## CHAPTER 6

Barry, R., Clarke, A. and Johnstone, S. (2003). A review of electrophysiology in attention-deficit/hyperactivity disorder: I. Qualitative and quantitative electroencephalography. *Clinical Neurophysiology* 114: 171–183.

Chalmbers, D. J. (1996). *The Conscious Mind: In Search of a Fundamental Theory*. Oxford University Press, New York.

Cheek, T. R. (1989). Spatial aspects of calcium signaling. *Journal of Cellular Science* 93: 211–216.

Chervin, R. D., Archibold, K. H., Dillon, J. E., Panahi, P., Pituch, K. J., Dahl, R. E. and Guilleminault, C. (2002). Inattention, hyperactivity, and symptoms of sleep disordered breathing. *Pediatrics* 109(3): 449–456.

Christakos, C. N. (1986). The mathematical basis of population rhythms in nervous and neuromuscular systems. *International Journal of Neuroscience* 29: 103–107.

Clark, A., Barry, R., McCarthy, R. and Selikowitz, M. (2001). EEG-defined subtypes of children with attention-deficit/hyperactivity disorder. *Clinical Neurophysiology* 112: 2098–2105.

Crick, F. and Mitchinson, G. (1983). The function of dream sleep. *Nature* 304: 111–114.

Domhoff, G. W. (2003). *The Scientific Study of Dreams; Neural Networks, Cognitive Development and Content Analysis*. American Psychological Association, Washington, D.C.

Foulkes, D. (1985). *Dreaming: A Cognitive–Psychological Analysis*. Lawrence Erlbaum Associates, Hillsdale, NJ.

Gilman, A. G. (1989). G proteins and the regulation of adenylyl cyclase. *Journal of the American Medical Association* 162: 1819–1825.

Goodenough, D. R., Lewis, H. B., Shapiro, A., Jaret, L. and Sleser, I. (1965). Dream reporting following abrupt and gradual awakenings from different types of sleep. *Journal of Personality and Social Psychology* 2: 170–179.

Harold, F. M. (1986). *The Vital Force: A Study of Bioenergetics*. WH Freeman and Co., New York.

Herrmann, W. M. and Schaerer, E. (1986). Pharmaco-EEG: Computer EEG analysis to describe the projection of drug effects on a functional cerebral level in humans. In *Clinical Application of Computer Analysis of EEG and Other Neurophysiological Signals, EEG Handbook* (Revised Series), Vol. 2, eds. Lopes et al. Elsevier, Amsterdam, pp. 385–445.

Hodgkin, A. L. and Horowicz, P. (1959). The influence of potassium and chloride ions on the membrane potential of single muscle fibers. *Journal of Physiology* 148: 127–160.

Itil, T. M. (1981). The discovery of psychotropic drugs by computer analyzed cerebral bio-electrical potentials (CEEG). *Drug Development and Research* 1: 373–407.

John, E. R. and Swartz, E. L. (1978). The neurophysiology of information processing and cognition. *Annual Review of Psychology* 29: 1–29.

Kahn, D. and Hobson, J. (2005). State-dependent thinking: A comparison of waking and dreaming thought. *Conscious Cognition* 14(3): 429.

Koukkou, M. and Lehmann, D. (1983). Dreaming: the functional state-shift hypothesis, a neuropsychophysiological model. *British Journal of Psychiatry* 142: 221–231.

Krebs, E. G. (1989). Role of the cycle AMP-dependent protein kinase in signal transduction. *Journal of the American Medical Association* 262: 1815–1818.

Lutz, A. (2006). Using meditation to study brain neuroplasticity and neural correlates of objective experience. *Towards a Science of Consciousness*, Center for Consciousness Studies, University of Arizona. Consciousness Research Abstracts. Tucson, AZ, pp. 177–178.

Mandema, J. W. and Danhof, M. (1992). Electroencephalogram effect measures and relationships between pharmacokinetics and pharmacodynamics of centrally acting drugs. *Clinical Pharmacokinetics* 23(3): 191–215.

Monstra, V., Lubar, J., Linden, M., VanDeusen, P., Green, G., Wing, W., Phillips, A. and Fenger, T. (1999). Assessing attention deficit hyperactivity disorder via quantitative electroencephalography: An initial validation study. *Neuropsychology* 13(3): 424–433.

Ogilvie, R. D., Hunt, H. T., Tyson, P. D., Lucescu, M. and Jenkins, D. B. (1982). Lucid dreaming and alpha activity: A preliminary report. *Perceptual and Motor skills* 55(3 Pt 1): 795–808.

Pagel, J. F. (1990). Proposing an electrophysiology for state dependent sleep and dream mentation. *APSS Abstracts*, 134.

Pagel, J. F. (1993a). Neurosignalling: Electrophysiologic volume conduction in the CNS. *International Conference on the Cellular Consequences of Sleep* (Abstracts), Mawi, Howaii, p. 606.

Pagel, J. F. (1993b). Modeling drug actions on electrophysiologic effects produced by EEG modulated potentials. *Human Psychopharmachology* 8(3): 211–216.

Pagel, J. F. (1994). Modeling drug actions on active EEG effects. *APSS Abstracts*. p. 212.

Pagel, J. F. (1996). Pharmachologic alterations of sleep and dream: A clinical framework for utilizing the electrophysiological and sleep stage effects of psychoactive medications. *Human Psychopharmachology* (11): 217–223.

Pagel, J. F. (2004). Drug induced alterations in dreaming: An exploration of the dream data terrain outside activation–synthesis. *Behavior and Brain Science* 27(5): 10–14.

Pagel, J. F. (2005). Neurosignals – Incorporating CNS electrophysiology into cognitive process. *Behavioral and Brain Sciences* 28(1): 75–76.

Pagel, J. F. and Pandi-Perumal, S. R. (eds.) (2007). *Primary Care Sleep Medicine – A Practical Guide*. Humana Press, Totowa, NJ.

Pagel, J. F. and Shocknesse, S. (2005). *EEG Alpha Intrusion on EEG is Not Associated with Changes in Dream or Nightmare Frequency*. Data from Sleep Disorders Center of Southern Colorado, Pueblo, Colorado.

Pagel, J. F., Snyder, S. and Dawson, D. (2004). Obstructive sleep apnea in sleepy pediatric psychiatry clinic patients: Polysomnographic and clinical correlates. *Sleep and Breathing* 8(3): 125–131.

Pagel, J. F., Dawson, D., Drozd, J. and Coolidge, F. (2006a). Personality and neuropsychological characteristics identifying children with OSA in a pediatric AD/HD population. *Sleep Abstracts*.

Pagel, J. F., Dawson, D., Drozd, J. and Snyder, S. (2006b). Quantitative EEG (QEEG) power characteristics of children with OSA in a pediatric AD/HD population. *Sleep* 29(Suppl): A337–A338.

Pagel, J. F. and Snyder, S. (2007). Obstructive sleep apnea associated QEEG abnormalities in pediatric patients with QEEG based diagnosis of AD/HD. *Sleep* (30) Abstract Supplement. p. A341.

Plato, Plato's Cave – the Second Stage, from The Republic, book VII, 514a–517a. taken from Platonis Oprea, recogn. Ioannes Burnet, Oxonii: clarendon, 2nd Edition (1905–1910), Vol. 4 (English trans.) M. Heidegger included in The Essence of Truth (2002) Continuum, London.

Siegel, J. M. (2000). Brainstem mechanisms generating REM sleep. In *Principles and Practice of Sleep Medicine*, 3rd Edition, eds. M. H. Kryger, T. Roth and W. C. Dement. W. B. Saunders Co., Philadelphia, PA, pp. 112–114.

Steriade, M. (2001). *The Intact and Sliced Brain*. The MIT Press, Cambridge, MA.

CHAPTER 7

Barrie, J. (1911/1988). *Peter Pan and Wendy*. Clarkson N. Potter, Inc./Publishers (Crown), New York.

Bliss, T. V. and Lomo, T. (1973). Long lasting potentiation of synaptic transmission in the dentate area of the anesthesized rabbit following stimulation of the perforant path. *Journal of Physiology* 232: 331–356.

Coveny, P. and Highfield, R. (1995). *Frontiers of Complexity*. Fawcett Columbine, New York.

Crick, F. and Mitchinson, G. (1983). The function of dream sleep. *Nature* 304: 111–114.

Crick, F. and Mitchinson, G. (1995). REM sleep and neural nets. *Behavioral Brain Research* 69: 147–155.

Davies, P. (2006). Advances in cosmology suggest a link between information, complexity and the age of the universe. This development could remove a fundamental obstacle to strong emergence in nature. *Towards A Science of Consciousness*, Center for Consciousness Studies, University of Arizona, Tucson, AZ, pp. 152–153.

Hebb, D. (1949). *Organization of Behavior*. Wiley, New York, p. 62.

Hinton, G., Sejnowski, T., Ackey, D. and Boltzmann (1984). Machines: Constant satisfaction networks that learn, Technical Report CMU CS 84, 111. Carnegie-Mellon University, Pittsburg.

McCullough, W. S. and Pitts, W. (1943). *Bulletin of Mathematical Biophysics* 5: 115.

Traub, R., Miles, R. and Buzsaki, G. (1992). Computer simulation of carbachol-driven rhythmic population oscillations in the CA3 region of the in vitro rat hippocampus. *J. Physiol* 451: 653–672.

SECTION 3

Cavallero, C., Foulkes, D., Hollifield, M. and Terry, R. (1990). Memory sources of REM an NREM dreams. *Sleep* 13: 449–455.

Ekstrand, B., Barrett, T., West, J. and Maier, W. (1977). The effects of sleep on long term memory. In *Neurobiology of Sleep and Memory*, eds. R. Drucker-Colin and J. McGaugh. Academic Press, Philadelphia.

Kuiken, D. and Sikora, S. (1993). The impact of dreams on waking thoughts and feelings. In *The Functions of Dreaming*, eds. A. Moffitt, M. Kramer and R. Hoffman. State University of New York Press, Albany, NY, pp. 419–476.

Moffitt, A., Kramer, M. and Hoffmann, R. (1993). *The Functions of Dreaming*. State University of New York Press, Albany, NY. pp. 419–476.

Pagel, J. F. and Vann, B. (1992). The effects of dreaming on awake behavior. *Dreaming* 2(4): 229–237.

Pagel, J. F. and Vann, B. (1993). Cross-cultural dream use in Hawaii. *Hawaii Medical Journal* 52(2): 44–45.

Pagel, J. F. and Vann, B. (1995). Polysomnographic correlates of reported dreaming: Negative correlation of RDI with dreaming effects on waking activity. *APSS Abstracts*. p. 51.

Schredl, M. (2002). Questionnaires and diaries as research instruments in dream research: Methodological issues. *Dreaming* 12(1): 17–26.

CHAPTER 8

Armitage, R., Rochlen, A., Fitch, T., Trivedi, M. and Rush, A. (1995). Dream recall and major depression: A preliminary report. *Dreaming* 5: 189–198.

Belicki, K., Chambers, E. and Ogilvie, R. (1997). Sleep quality and nightmares. *Sleep Research* 26: 637.

Bernert, R., Joiner, T., Cukrowicz, K., Schmidt, N. and Krakow, B. (2005). Suicidality and sleep disturbances. *Sleep* 28(9): 1135–1141.

Besiroglu, L., Agargun, M. and Inci, R. (2005). Nightmares and terminal insomnia in depressed patients with and without melancholic features. *Psychiatry Research* 133: 285–287.

Breslau, N., Roth, T., Rosenthal, L. and Andreski, P. (1996). Sleep disturbance and psychiatric disorders: A longitudinal epidemiological study of young adults. *Biological Psychiatry* 39(6): 411–418.

Chervin, R. (2003). Use of clinical tools and tests in sleep medicine. In *Principles and Practice of Sleep Medicine*, 3rd Edition, eds. M. Kryger, T. Roth and W. Dement. W. B. Saunders Co., Philadelphia, PA, pp. 535–546.

Edinger, J. (Chair); Work group members: Bonnet, M., Bootzin, R., Doghramp, K., Dorsey, C., Espie, C., Jamieson, A., McCall, W., Morin, C. and Stepanski, E. (2004). Derivation of research diagnostic criteria for insomnia: Report of an American Academy of Sleep Medicine work group. *Sleep* 27(8): 1567–1592.

Freud, S. (1917). Difficulties and first approaches, general theory of the neuroses. In *Introductory Lectures on Psychoanalysis (1966)*, trans. and ed. J. Strachey. W. W. Norton, New York, p. 89.

Hohagen, F., Rink, H., Kappler, C., Schramm, E., Riemann, D., Weyerer, S. and Berger, M. (1993). Prevalence and treatment of insomnia in general practice: A longitudinal study. *European Archives of Psychiatry Clinic Neuroscience* 242: 326–329.

Knipling, R. R. and Wang, S. S. (1995). Revised estimates of the US drowsy driver crash problem size based on General Estimates System case reviews. *39th Annual Proceedings, Association for the Advancement of Automotive Medicine*, Chicago, IL.

Krakow, B., Germain, A. and Warner, T. (2001). The relationship of sleep quality and post-traumatic stress to potential sleep disorders in sexual assault survivors with nightmares insomnia and PTSD. *Journal of Traumatic Stress* 14: 647–665.

Kuiken, D. and Sikora, S. (1993). The impact of dreams on waking thoughts and feelings. In *The Functions of Dreaming*, eds. A. Moffitt, M. Kramer and R. Hoffman. State University of New York Press, Albany, NY, pp. 419–476.

Leister, A. (Chairperson) (2005). National Institutes of Health state of the science conference statement: Manifestations and management of chronic insomnia in adults June 13–15, 2005. *Sleep* 28(9): 1049–1057.

Levin, R. (1994). Sleep and dreaming characteristics of frequent nightmare subjects in a university population. *Dreaming* 4(2): 127–138.

Myers, P. and Pagel, J. F.(2001). The effects of daytime sleepiness and sleep onset REMS period (SORP) on reported dream recall. *Sleep* 24: A183.

Ohayon, M. (2002). Epidemiology of insomnia: What we know and what we still need to learn. *Sleep Medicine Review* 6(2): 97–111.

Ohayon, M. and Morselli, N. (1994). Nightmares: Their relation with insomnia, mental illness and diurnal functioning. *Sleep Research* 23: 171.

Pagel, J. F. (1994). The treatment of insomnia. *American Family Physician* 49(6): 1417–1422.

Pagel, J. F. and Nielsen, T. (2005). Parasomnias: Recurrent Nightmares – American Academy of Sleep Medicine. The International Classification of Sleep Disorders, revised: Diagnostic and Coding Manual (ICD-10). American Academy of Sleep Medicine, Chicago, IL.

Pagel, J. F. and Shocknesse, S. (2006). The Sleep Substrate of Dreaming: Polysomnographic correlates of reported dream recall frequency, *Towards a Science of Consciousness*, Tucson, AZ.

Pagel, J. F. and Vann, B. (1995). Polysomnographic correlates of reported dreaming: Negative correlation of RDI with dreaming effects on waking activity. *APSS Abstracts*. p. 51.

Revonsuo, A. (2003). The reinterpretation of dreams: An evolutionary hypothesis of the function of dreaming. In *Sleep and Dreaming: Scientific Advances and Reconsiderations*, eds. E. Pace-Schott, M. Solms, M. Blagtove and S. Harnad. pp. 85–111.

Schredl, M. (2001). Dream recall frequency and sleep quality of patients with restless legs syndrome. *European Journal of Neurology* 8: 185–189.

Shakespeare, W. Hamlet III,I,36. (1975). *The Complete Works of William Shakespeare*. Gramercy Books, New York, p. 1088.

Solms, M. (1997). *The Neuropsychology of Dreams – A Clinico-anatomical Study*. Lawrence Erlbaum Associates, Mahwah, NJ.

Wood, J. and Bootzin, R. (1990). The prevalence of nightmares and their independence from anxiety. *Journal of Abnormal Psychology* 99: 64–68.

CHAPTER 9

Armitage, R., Rochlen, A., Fitch, T., Trivedi, M. and Rush, A. (1995). Dream recall and major depression: A preliminary report. *Dreaming* 5: 189–198.

Blagrove, M. and Akehurst, L. (2000). Personality and dream recall frequency: Further negative findings. *Dreaming* 10(3): 139–148.

Cavallero, C. and Foulkes, D. (1993). *Dreaming as Cognition*. Harverter-Wheatsheaf, New York.

Cohen, D. B. (1979). *Sleep and Dreaming: Origins, Nature and Function*. Pergamon, New York.

Cohen, G. (1996). *Memory in the Real world*, 2nd Edition. Psychology Press, Hove, East Sussex, pp. 286–287.

DeWitt, T. (1998). Impairment of reality constructing processes in dream experience. *Journal of Mental Imagery* 12: 65–78.

Domhoff, G. (2003). *The Scientific Study of Dreams: Neural Networks, Cognitive Development and Content Analysis*. American Psychological Association, Washington, D.C.

Domhoff, G. and Schneider, A. (1998). New rationales and methods for quantitative dream research outside the laboratory. *Sleep* 12: 398–404.

Duffy, J. D. and Campbell, J. J. (1994). The regional pre-frontal syndromes: A theoretical and clinical overview. *Journal of Neuropsychiatry* 6(4): 379–387.

Eich, E. (1989). Theoretical issues in state dependent memory. In *Varieties of Memory and Consciousness*, eds. H. Roediger and F. Craik. Lawrence Erlbaum Associates, Hillsdale, NJ, pp. 331–352.

Ekstrand, B. (1972). To sleep, perchance to dream (about why we forget). In *Human Memory – Festschrift in Honor of Benton J. Underwood*, eds. C. Duncan, L. Secherest and A. Melton. Appleton-Century-Crofts, New York, pp. 59–82.

Fogel, B. S. (1994). The significance of frontal system disorders for medical practice and health policy. *Journal of Neuropsychiatry* 6(4): 343–347.

Foulkes, D. (1985). *Dreaming: A Cognitive–Psychological Analysis*. Lawrence Erlbaum Associates, Hillsdale, NJ.

Garfield, P. (1991). *The Healing Power of Dreams*. Simon and Schuster, New York.

Giambra, L. (1997). Sex differences in daydream and related mental activity from the late teens to the early nineties. *International Journal of Aging and Human Development* 10: 1–34.

Goodenough, D. R. (1991). Dream recall: History and current status of the field. In *The Mind in Sleep*, eds. S. J. Ellman and J. S. Antrobus. John Wiley and Sons, New York.

Goodenough, D. R., Lewis, H. B., Shapiro, A., Jaret, L. and Sleser, I. (1965). Dream reporting following abrupt and gradual awakenings from different types of sleep. *Journal of Personality and Social Psychology* 2: 170–179.

Hall, C. S. and Van de Castle, R. L. (1966). *The Content Analysis of Dreams*. Appleton-Century-Crofts, New York.

Hartmann, E. (1998). *Dreams and Nightmares – The New Theory on the Origin and Meaning of Dreams*. Plenum Trade, New York and London.

Kosslyn, S. M. (1994). *Image and Brain – The Resolution of the Imagery Debate, A Bradford Book*. The MIT Press, Cambridge, MA.

Koukkou, M. and Lehmann, D. (1983). Dreaming: the functional state-shift hypothesis, a neuropsychophysiological model. *British Journal of Psychiatry* 142: 221–231.

Kramer, M. (1993). The selective mood regulatory function of dreaming: An update and revision. In *The Functions of Dreaming*, eds. A. Moffitt, M. Kramer and R. Hoffman. State University of New York Press, Albany, NY, pp. 139–196.

Kramer, M. and Roth, T. (1979). Dreams in psychopathology. In *Handbook of Dreams: Research, Theories and Applications*, ed. B. Wolman. Von Norstrand Reinhold, New York, pp. 361–387.

Kuiken, D. and Sikora, S. (1993). The impact of dreams on waking thoughts and feelings. In *The Functions of Dreaming*, eds. A. Moffitt, M. Kramer and R. Hoffman. State University of New York Press, Albany, NY, pp. 419–476.

Marks, D. (1990). On the relationship between imagery, body and mind. In *Imagery – Current Developments*, eds. P. Hampson, D. Marks and J. Richardson. Routledge, New York, pp. 1–36.

McCarley, R. and Hobson, J. (1975). Neuronal excitability modulation over the sleep cycle: A structural and mathematical model. *Science* 189: 58–60.

Moscovitch, M. (1989). Confabulation and the frontal systems: Strategic versus associative retrieval in neuropsychological theories of memory. In *Varieties of Memory and Consciousness*, eds. H. Roediger and F. Craik. Lawrence Erlbaum Associates, Hillsdale, NJ, pp. 133–159.

Myers, P. and Pagel, J. F. (2001). The effects of daytime sleepiness and sleep onset REMS period (SORP) on reported dream recall. *Sleep* 24: A183.

Pagel, J. F. and Vann, B. (1992). The effects of dreaming on awake behavior. *Dreaming* 2(4): 229–237.

Pagel, J. F. and Vann, B. (1993). Cross-cultural dream use in Hawaii. *Hawaii Medical Journal* 52(2): 44–45.

Pagel, J. F. and Vann, B. (1996). A learning/memory based paradigm describing cognitive integration of dreams: Evidence for such a paradigm derived from reported dream effects on awake behavior in obstructive sleep apnics. *ASD Abstracts*.

Pagel, J. F. and Vann, B. (1995). Polysomnographic correlates of reported dreaming: Negative correlation of RDI with dreaming effects on waking activity. *APSS Abstracts*.

Pagel, J. F., Vann, B. and Altomare, C. (1995). The association between stress and dreaming: Community background levels and changes with disaster (Hurricane Iniki). *Dreaming* 5(1): 43–50.

Renya, V. (1995). Interference effects in memory and reasoning, a fuzzy-trace theory analysis. In *Interference and Inhibition in Cognition*, eds. F. Dempsater and C. Brainerd. Academic Press Inc., New York, pp. 29–59.

Roediger, H., Weldon, M. and Challis, B. (1989). Explaining dissociations between implicit and explicit measures of retention: A processing account. In *Varieties of Memory and Consciousness*, eds. H. Roediger and F. Craik. Lawrence Erlbaum Associates, Hillsdale, NJ, pp. 3–41.

Solms, M. (1997). *The Neuropsychology of Dreams – A Clinico-Anatomical Study*. Lawrence Erlbaum Associates, Mahwah, NJ.

Spear, N. and Riccio, D. (1994). *Memory – Phenomena and Principles*. Allyn and Bacon, Boston, MA, pp. 66–107.

States, B. (1997). *Seeing in the Dark: Reflections on Dreams and Dreaming*. Yale University Press, New Haven, p. 215.

States, B. O. (1993). Bizarrness in dreams and fictions. In *The Dream and the Text*, ed. C. S. Rupprecht. State University of New York Press, Albany, NY.

Strauch, I. and Meier, B. (1996). *In Search of Dreams: Results of Experimental Dream Research*. State University of New York Press, Albany, NY.

Tedlock, B. (ed.) (1992). *Dreaming: Anthropological and Psychological Interpretations*. School of American Research Advance Seminar Series, Sante Fe, New Mexico.

Tulving, E. (1983). *Elements of Episodic Memory*. Oxford University Press, New York.

Wilde, O. (1895/1941). The importance of being ernest. In *The Literature of England*, Vol. 2, 3rd Edition, eds. G. Woods, H. Watt and Aderson G. Scott, Forsman, Chicago, IL.

Wolcott, S. and Strapp, C. (2002). Dream recall frequency and dream detail as mediated by personality behavior and attitude. *Dreaming* 12(1): 27–44.

CHAPTER 10

Benson, D. (1994). *The Neurology of Thinking*. Oxford University Press, New York, p. 14.

Calvin, W. (1996). *How Brains Think: Evolving Intelligence, Then and Now*. Basic Books, New York.

Caudill, M. (1992). *In Our Own Image – Building an Artificial Person*. Oxford University Press, Oxford. p. 214.

Crick, F. and Mitchinson, G. (1983). The function of dream sleep. *Nature* 304: 111–114.

Crick, F. and Mitchinson, G. (1995). REM sleep and neural nets. *Behavioral Brain Research* 69: 147–155.

Davies, P. (2006). Advances in cosmology suggest a link between information, complexity and the age of the universe. This development could remove a fundamental obstacle to strong emergence in nature. Center for Consciousness Studies, University of Arizona, Consciousness Research Abstracts. *Towards A Science of Consciousness*, Tucson, Arizona, 152–153.

Gall, F. (1938). *On the Function of the Cerebellum*. Machlachlan and Stewart, Edinburgh.

Gould, J. and Gould, C. (1994). The animal mind. W. H. Freeman, New York.

Lenhert, W. and Ringle, M. (1982). *Strategies for Natural Language Processing*. Erlbaum and Associates, Hillsdale, NJ.

Newell, A. and Simon, H. (1972). *Human Problem Solving*. Prentice-Hall, Englewood Cliffs, NJ.

Pearl, J. (1988). *Probabilistic Reasoning in Intelligent Systems*. Morgan-Kaufmann, San Mateo, CA.

Penrose, R. (1990). *The Emperors New Mind*. Oxford University Press, New York.

Pierce, J. (1980). *An Introduction to Information Theory: Symbols, Signals and Noise* 2nd revised Edition. Dover Publications, Inc., New York.

Pinker, S. (1994). *The Language Instinct*. Harper and Collins, New York.

Pogliano, C. (1991). Between form and function: A new science of man. In *The Enchanted Loom: Chapters in the History of Neuroscience*, ed. P. Corsi. Oxford University Press, Oxford.

Russel, B. (1927). *Philosophy*. Norton, New York.

Sampson, J. (1984). *Biological Information Processing: Current Theory and Computer Simulation*. John Wiley and Sons, New York.

CHAPTER 11

Amaral, D. (2002). The primate amygala and the neurobiology of social behavior: Implications for understanding social anxiety. *Biological Psychiatry* 51(1): 11–17.

Damasio, A. (2001). Fundamental feelings. *Nature* 413: 67, 781.

Damasio, A. (2003). *Looking for Spinoza – Joy, Sorrow, and the Feeling Brain.* Harcourt Inc., New York, pp. 5, 64, 84.

Damasio, A., Grabowski, T., Bechara, A., Damasio, H., Pronto, L., Parvizi, J. and Hichwa, R. (2001). Subcortical and cortical brain activity during the feeling of self generated emotions. *Nature Neuroscience* 4: 207–212.

Darwin, C. (1872/2005). *The Expression of Emotions in Man and Animals in Darwin: The Indelible Stamp*, ed. J. Watson. Running Press, Philadelphia, PA, pp. 1066–1258.

Ekstrand, B., Barrett, T., West, J. and Maier, W. (1977). The effect of sleep on human long-term memory. In *Neurobiology of Sleep and Memory*, eds. R. Drucker-Colin and J. McGaugh. Academic Press, New York, pp. 419–438.

Francis, A. (Chair – Task Force on DSM-IV) (1994). *DSM-IV – Diagnostic and Statistical Manual of Mental Disorders*, 4th Edition. American Psychiatric Association, Washington, D.C.

Freud, S. (1916/1951). *Beyond the Pleasure Principle*, Vol. 18, the Standard Edition, trans. J. Strachey. Hogarth, London.

Hartmann, E. (1994). Nightmares and other dreams. In *Principles and Practice of Sleep Medicine*, 2nd Edition, eds. M. Kryger, T. Roth and W. Dement. W. B. Saunders, London, pp. 407–410.

Kramer, M. (1993). The selective mood regulatory function of dreaming: An update and revision. In *The Functions of Dreaming*, eds. A. Moffitt, M. Kramer and R. Hoffman. State University of New York Press, Albany, NY, pp. 139–195.

Levin, R. and Nielsen, T. (2006). Nightmares, PTSD and affect distress: A proposed neurocognitive model of nightmare production. In *Towards a Science of Consciousness 2006.* pp. 122–123. Accepted for publication in full format Psychology Bulletin 2007.

Pagel, J. F. and Nielsen, T. (2005). Parasomnias: Recurrent Nightmares – American Academy of Sleep Medicine. The International Classification of Sleep Disorders, revised: Diagnostic and Coding Manual (ICD-10). American Academy of Sleep Medicine, Chicago, IL.

Pagel, J. F., Kwiakowski, C. and Broyles, K. (1999). Dream use in filmmaking. *Sleep* 22(Suppl): S173.

Piccione, P., Jacobs, G., Kramer, M. and Roth, T. (1977). The relationship between daily activities, emotions and dream content. *Sleep Research* 6: 133.

Shakespeare, W. Hamlet, ACT II. Sc. 2 (1975). In *The Complete Works of William Shakespeare*. Gramercy Books, New York.

CHAPTER 12

Cavell, S. (1971). *The World Viewed – Reflections on the Ontology of Film.* The Viking Press, New York.

Coleridge, S. (1912/1979). *Coleridge: Poetical Works*, ed. E. Coleridge. Oxford University Press, Oxford.

Collier, J. and Collier, M. (1912/1986). *Visual Anthropology: Photography as a Research Method.* University of New Mexico Press, Albuquerque.

Damasio, A. (1994). *Descartes' Error – Emotion, Reason and the Human Brain.* Avon Books Inc., New York.

Hall, E. (1986). Foreword. In *Visual Anthropology: Photography as a Research Method (1912/1986)*, eds. J. Collier and M. Collier. University of New Mexico Press, Albuquerque, p. xvi.

Hetch, H., Swartz, R. and Atherton, M. (eds.) (2003). *Looking into Pictures: An Interdisciplinary Approach to Pictorial Space.* The MIT Press, Cambridge, MA, p. xii.

Hopkins, R. (2003). Perspective, convention, and compromise. In *Looking into Pictures: An Interdisciplinary Approach to Pictorial Space*, eds. H. Hetch, R. Swartz and M. Atherton. The MIT Press, Cambridge, MA, pp. 145–166.

Klatsky, R. and Lederman, S. (1993). Spatial and non-spatial avenues to object recognition by the human haptic system. In *Philosophy and Psychology*, eds. N. Eilian, R. McCarthy and B. Brewer. Oxford University Press, Oxford, pp. 191–205.

Kosslyn, S. M. (1994). *Image and Brain – The Resolution of the Imagery Debate, A Bradford Book*. The MIT Press, Cambridge, MA.

LaBerge, S. (1995). *Lucid Dreaming: The Power of Being Awake and Aware in Your Dreams*. Ballantine Books, New York.

Leger, F. (1938). *fonctions de la peinture*. Gothier, Paris.

Marks, D. (1990). On the relationship between imagery, body and mind. In *Imagery – Current Developments*, eds. P. Hampson, D. Marks and J. Richardson. Routledge, New York, pp. 1–36.

Mausfield, R. (2003). Conjoint representations. In *Looking into Pictures: An Interdisciplinary Approach to Pictorial Space*, eds. H. Hetch, R. Swartz and M. Atherton. The MIT Press, Cambridge, MA, pp. 17–60.

McCarthy, R. (1993). Introduction: What and where, in spatial representation – problems. In *Philosophy and Psychology*, eds. N. Eilian, R. McCarthy and B. Brewer. Oxford University Press, Oxford, pp. 319–324.

O'Keefe, J. (1993). Kant and the sea-horse: An essay in the neurophilosophy of space, in spatial representation – problems. In *Philosophy and Psychology*, eds. N. Eilian, R. McCarthy and B. Brewer. Oxford University Press, Oxford, pp. 43–64.

Sachs-Hombach, K. (2003). Resemblance reconceived. In *Looking into Pictures: An Interdisciplinary Approach to Pictorial Space*, eds. H. Hetch, R. Swartz and M. Atherton. The MIT Press, Cambridge, MA, pp. 167–178.

Sartre, J. (1991). The imaginary life. In *The Psychology of Imagination*. Benard Frechtman, New York, pp. 177–212.

Scary, E. (1995). On vivacity: The difference between daydreaming and imagining-under-authorial-instruction. *Representations* 52: 1–26.

States, B. (1997). *Seeing in the Dark: Reflections on Dreams and Dreaming*. Yale University Press, New Haven, p. 97.

Treisman, A. (2000). The binding problem. In *Findings and Current Opinion in Cognitive Neuroscience*, eds. L. Squire and S. Kosslyn. The MIT Press, Cambridge, MA, pp. 31–38.

Vuilleumier, P. and Swartz, S. (2001). Beware and be aware: Capture of spatial attention by fear-related stimuli in neglect. *NeuroReport* 12: 1119–1122.

Zeki, S. (1999). *Inner Vision: An Exploration of Art and the Brain*. Oxford University Press, Oxford, pp. 66–67.

CHAPTER 13

Arieti, S. (1976). *Creativity – The Magic Synthesis*. Basic Books, New York.

Belth, M. (1977). *The Process of Thinking*. David McKay Co., New York, p. xvii.

Benson, D. (1994). *The Neurology of Thinking*. Oxford University Press, New York, pp. 26, 206, 223, 239, 243.

Bonato, R., Moffitt, R., Hoffamnn, M., Cuddy, M. and Wimmer, F. (1991). *Dreaming* 1: 53–61.

Carter, R. (1998). *Mapping the Brain*. University of California Press, Berkeley, CA, p. 194.

Flanagan, O. (2000). *Dreaming Souls: Sleep, Dreams and the Evolution of the Conscious Mind*. Oxford University Press, New York, p. 148.

Fosse, R., Stickgold, R. and Hobson, J. (2001). Brain-mind states: reciprocal variation in thoughts and hallucinations. *Psychological Science* 12(1): 30–36.

Hartmann, E. (1998). *Dream and Nightmares: The New theory for the Origin and Meaning of Dreams*. Plenum Press, New York, p. 226.

Heidegger, M. (1968). *What is Called Thinking*, trans. J. Glenn Gray. Harper and Row Publishers, New York, p. xv.

Hunt, H. (1989). *The Multiplicity of Dreams: Memory, Imagination, Consciousness*. Yale University Press, New Haven.

Hunt, H. (1991). Dreams as literature/science: An essay. *Dreaming* 1: 235–242.

Jackson, J. (1932). *Selected Writings of J. Hughlings Jackson*, ed. J. Taylor. Hoddler and Stoughton, London.

Kahn, D. and Hobson, J. (2005). State-dependent thinking: A comparison of waking and dreaming thought. *Conscious Cognition* 14(3): p. 429.

Kahn, D. and Hobson, J. (2006). *State-Dependent Thinking: A Comparison of Waking and Dreaming Thought. Towards a Science of Consciousness 2006*, Center for Consciousness Studies, University of Arizona, Tucson, AZ, p. 121.

Lutz, J. (2006). *Using Meditation Expertise to Study Brain Neuroplasticity and the Neural Correlates of Subjective Experience. Towards a Science of Consciousness 2006*, Center for Consciousness Studies, University of Arizona, Tucson, AZ, p. 177.

Mamelak, A. and Hobson, J. (1989). Dream bizarreness as the cognitive correlate of altered neuronal behavior in REM sleep. *Journal of Cognitive Neuroscience* 1: 201–222.

Mesulam, M. (1985). Attention, confusional states and neglect. In *Principles of Behavioral Neurology*, ed. M. Mesulam. F. A. Davis, Philadelphia, PA, pp. 125–168.

Overstreet, D. (1980). Oxymoronic language and logic in quantum mechanics and James Joyce. *Sub-Stance* 28: 37–59.

Rittenhouse, C., Stickgold, R. and Hobson, J. (1994). Constraints on the transformation of characters, objects, and settings in dream reports. *Consciousness and Cognition* 3: 100–113.

Sartre, J. (1968). *The Psychology of Imagination*, trans. B. Frechtman. Washington Square Press, New York, p. 216.

Shoben, E. J. (1961). Culture ego psychology, and an image of man. *American Journal of Psychotherapy* 15: 395–408.

Stanovich, K. and West, R. (2000). Individual differences in reasoning: Implications for the rationality debate. *Behavioral and Brain Science* 23: 645.

States, B. (1993). *Dreaming and Storytelling*. Cornell University Press, Ithaca, NY. pp. 57, 99.

States, B. (1997). *Seeing in the Dark: Reflections on Dreams and Dreaming*. Yale University Press, New Haven, p. 224.

States, B. (2000). Dream bizarreness and inner thought. *Dreaming* 10(4): 26, 179–192.

States, B. O. (1993). Bizarrness in dreams and fictions. In *The Dream and the Text*, ed. C. S. Rupprecht. State University of New York Press, Albany, NY.

The Oxford English Dictionary (1971). Oxford University Press, New York.

Thomas, L. (1992). *The Fragile Species*. Charles Scribner and Sons, New York, p. 111.

Wolman, R. and Kozmova, M. (2006). Last night I had the strangest dream: Varieties of rational thought processes in dream report. *Consciousness and Cognition* di:10.1016/j.concog2006.09.009. Last accessed: Dec. 8. Accession Number: 17158070.

SECTION 5

Damasio, A. (1994). *Descartes' Error: Emotion, Reason and the Human Brain*. Avon Books Inc., New York.

Hobson, J. (1994). *The Chemistry of Conscious States – How the Brain Changes Its Mind*. Little, Brown, and Co., Boston, MA, p. 205.

CHAPTER 14

Asaad, G. and Shapiro, B. (1986). Hallucinations: theoretical and clinical overview. *American Journal of Psychiatry* 143: 1088–1097.

Black, D., Yates, W. and Andreasen, N. (1988). Schizophrenia, schizophreniform disorder and delusional (paranoid) disorders. In *Textbook of Psychiatry*, eds. J. Talbott, R. Hales and S. Yudofsky. The American Psychiatric Press, Washington, D.C., pp. 364–365.

Crick, F. and Mitchinson, G. (1983). The function of dream sleep. *Nature* 304: 111–114.

DSM-IV (1994). *Diagnostic and Statistical Manual of Mental Disorders*, 4th Edition, Francis A. (Chair – Task Force on DSM-IV). American Psychiatric Association, Washington, D.C.

Eliade, M. (1964). *Shamanism: Archaic Techniques of Ecstasy*, trans. W. Trask. Princeton University Press, Princeton, NJ, p. 503.

Esquirol, J. (1994). Le geste de Pinel: Psychiatric myth. In *Discovering the History of Psychiatry*, eds. M. Micale and R. Portor. Oxford University Press, Oxford, pp. 232–247.

Goldman, H. and Foreman, S. (2000). Psychopathology: Psychiatric diagnosis and psychosocial formulation. In *Review of General Psychiatry*, 5th Edition, ed. H. Goldman. Lange Medical Books/McGraw Hill, New York, p. 112.

Greenberg, D. and Brom, D. (2001). Nocturnal hallucinations in ultra-orthodox Jewish Israeli men. *Psychiatry* 64(1): 81–90.

Heron, W., Bexton, W. and Hebb, D. (1953). Cognitive effects of decreased variation to sensory environment. *American Psychologist* 8: 366.

Hobson, J. (1988). *The Dreaming Brain*. Basic Books Inc., New York, p. 210, 211.

Jaspers, K. (1923 trans.1962). *General Psychopathology*, trans. J. Hoenig and M. Hamilton. University Press. Manchester.

McKellar, P. (1957). *Imagination and Thinking*. Cohen and West, London.

Pagel, J. F. and Nielsen, T. (2005). *Parasomnias: Recurrent Nightmares – American Academy of Sleep Medicine. The International Classification of Sleep Disorders, Revised: Diagnostic and Coding Manual (ICD-10)*. American Academy of Sleep Medicine, Chicago, IL.

Perry, S. and Markowitz, J. (1988). Organic mental disorders. In *Textbook of Psychiatry*, eds. J. Talbott, R. Hales and S. Yudofsky. The American Psychiatric Press, Washington, D.C., p. 299.

Reed, G. (1972). *The Psychology of Anomalous Experience*. Houghton Miffen Co., Boston, MA, pp. 46, 48–53, 61–62, 67.

Shakespeare, W. (1986/2005). *MacBeth in Shakespeare – The Complete works*, 2nd Edition, eds. J. Jowett, W. Montogomery, G. Taylor and S. Wells. Clanrenden Press, Oxford.

CHAPTER 15

Aristotle (trans. 1984). On the soul (De Anima). In *Complete Works of Aristotle*, ed. J. Barnes, Revised Oxford Translation, Princeton 1: 687.

Ashbery, J. (1979). *Tapestry, in As We Know*. Viking, New York, p. 90.

Ashbery, J. (1985). And Ut Pictura Poesis is her name. In *Selected Poems*. Penguin Books, New York, p. 235.

Brannigan, E. (1992). *Narrative Comprehension and Film*. Routledge, London and New York, pp. 3, 4.

Foulkes, D. (1985). *Dreaming: A Cognitive–Psychological Analysis*. Lawrence Erlbaum Associates, Hillsdale, NJ.

Freud, S. (1914). Remebering, repeating and working-through. In *The Standard Edition of the Complete Psychological Works*, Vol. V, ed. J. Strachey (1973). Hogarth Press, London, p. 511.

Garfield, P. (1991). *The Healing Power of Dreams*. Simon and Schusteer, New York, p. 26.

Gibson, J. (1966). *The Senses Considered as Perceptual Systems*. Boston, MA, pp. 203–204.

Hardy, T. (1870). The Sun on the bookcase (students love song). In *Complete Poems of Thomas Hardy*, ed. J. Gibson. Palgrave-Macmillan, New York, p. 311.

Hardy, T. (reprinted 1985). *Tess of the d'Urbevvilles*. New York University Press, New York, p. 43.

Hunt, H. (1991). Dreams as literature/science: An essay. *Dreaming* 1: 235–242.

Kaplan, A. (ed.) (1990). *Psychoanalysis and Cinema*. Routledge, New York, p. 7, 11.

Koestler, A. (1969). *The Act of Creation*. Macmillan, New York, p. 45.

Mahoney, P. (1987). *Freud as a Writer*. Yale University Press, New Haven and London.

Mehring, M. (1990). *The Screenplay – A Blend of Film Form and Content*. Focal Press, Boston, MA.

Metz, C. (1982). *Psychoanalysis and Cinema*. Macmillan, New York.

Pagel, J. F., Kwiatkowski, C. and Broyles, K. (1999). Dream use in film making. *Dreaming* 9(4): 247–256.

Porter, L. M. (1993). Real dreams, literary dreams, and the fantastic in literature. In *The Dream and the Text – Essays on Literature and Language*, ed. C. S. Rupprecht. State University of New York Press, Albany, NY.

Potter, C. (1990). *Image, Sound and Story – The Art of Telling in Film*. Secker and Warburg, London.

Ricoeur, P. (1984a). *Time and Narrative*, Vol. 1. University of Chicago Press, Chicago, IL, p. 150.

Ricoeur, P. (1984b). *Time and Narrative*, Vol. 2. University of Chicago Press, Chicago, IL.

Sartre, J. (1991). The imaginary life, in the psychology of imagination. New. In *The Complete Works of William Shakespeare*. Gramercy Books, New York, 1975, p. 177.

Scarry, E. (1995). On vivacity: The difference between daydreaming and imagining-under-authorial-instruction. *Representations* 52: 1–26.

Silverman, K. (1983). *The Subject of Semiotics*. Oxford University Press, New York, p. 212.

States, B. (1992). *Hamlet and the Concept of Character*. Johns Hopkins University Press, Baltimore.

States, B. O. (1993). *Dreaming and Storytelling*. Cornell University Press, Ithaca, NY. p. 53.

States, B. (1994). Authorship in dreams and fictions. *Dreaming* 4(4): 237–253.

States, B. (1997). *Seeing in the Dark: Reflections on Dreams and Dreaming*. Yale University Press, New Haven, p. 206.

Stockholder, K. (1993). Dreaming of death: Love and money in the merchant of Venice. In *The Dream and the Text – Essays on Literature and Language*, ed. C. S. Rupprecht. State University of New York Press, Albany, NY.

Tedlock, B. (ed.) (1987). *Dreaming: Anthropological and Psychological Interpretations*. Cambridge University Press, Cambridge.

Todorov, T. (1981). *Introduction to Poetics*, trans. R. Howard. University of Minnesota Press, Minneapolis.

Vogler, C. (1992). *The Writer's Journey: Mythic Structure for Storytellers and Screenwriters*. A Michael Wiese Productions Book, Studio City, CA.

Westlund, J. (1993). Self and self-validation in a stage character: A Shakespearean use of dreams. In *The Dream and the Text – Essays on Literature and Language*, ed. C. S.Rupprecht. State University of New York Press, Albany, NY.

CHAPTER 16

Allen, R. (1995). *Projecting Illusion – Film Spectatorship and the Impression of Reality*. Cambridge University Press, Cambridge, UK, pp. 62, 115.

Baudry, J. (1986). The apparatus. In *Narrative, Apparatus, Ideology*, ed. P. Rosen. Columbia University Press, New York, pp. 236–318.

Bonitzer, P. (1992). Hitchcockian suspense. In *Everything You Always Wanted to Know About Lacan (But Were Afraid to Ask Hitchcock)*, ed. S. Zizek. Verso, London, pp. 15, 17.

Bulkeley, K. (1996). *Among All Those Dreamers: Essays on Dreaming in Modern Society*. State University of New York Press, Albany, NY.

Cartwright, R., Bernick, N., Borowitz, G. and Kling, A. (1969). Effects of an erotic movie on the dreams of young men. *Archives of General Psychiatry* 20: 263–271.

Chanan, M. (1996). *The Dream That Kicks: The Prehistory and Early Years of Cinema in Britain*. Routledge, London.

Cristie, I. (1994). *The Last Machine: Early Cinema and the Birth of the Modern World*. British Film Institute, London.

Dayan, D. (1976). The tutor code of classical cinema. In *Movies and Methods: An Anthology*, ed. B. Nichols. University of California Press, Berkeley, CA, pp. 438–451.

Eilen, N., McCarthy, R. and Brewer, B. (eds.) (1993). *Spatial Representation: Problems in Philosophy and Psychology*. Oxford University Press, Oxford.

Freud, S. (1914). Remembering, repeating and working-through. In *The Standard Edition of the Complete Psychological Works*, Vol. V, ed. J. Strachey (1973). Hogarth Press, London, pp. 536–537.

Friedberg, A. (1990). A denial of difference: Theories of cinematic Identification. In *Psychoanalysis and Cinema*, ed. E. Kaplan. Routledge, New York, p. 36.

Ibid., p. 44.

Gabbard, G. and Gabbard, K. (1999). *Psychiatry and the Cinema*, 2nd Edition. American Psychiatric Press, Washington, D.C., p. 194.

Hunt, H. (1989). *The Multiplicity of Dreams: Memory, Imagination and Consciousness*. Yale University Press, New Haven.

Kaplan, E. A. (ed.) (1990). *Psychoanalysis and Cinema*. Routledge, New York, p. 10.

Klatsky, R. and Lederman, S. (1993). Spatial and non-spatial avenues to object recognition by the human haptic system. In *Philosophy and Psychology*, eds. N. Eilian, R. McCarthy and B. Brewer. Oxford University Press, Oxford, pp. 191–205.

Kosslyn, S. M. (1994). *Image and Brain – The Resolution of the Imagery Debate, A Bradford Book*. The MIT Press, Cambridge, MA.

Lacan, J. (1979a). Desire and the interpretation of desire in Hamlet. In *Literature and Psychoanalysis: The Question of Reading Otherwise*, ed. Feldman. Johns Hopkins University Press, pp. 11–52.

Lacan, J. (1979b). The imaginary signifier. In *Four fundamental Concepts of Psychoanalysis*, trans. A. Sheridan. Penguin Books Ltd., Hammondsworth, p. 75.

Olson, C. and Gettner, S. (2000). Brain representation of object-centered space. In *Findings and Current Opinion in Cognitive Neuroscience*, eds. L. Squire and S. Kosslyn. The MIT Press, Cambridge, MA, pp. 25–30.

Oudart, J. (1978). Cinema and suture. *Screen* 18(4): 35–47.

Passmore, J. (1991). *Serious Art*. Open Court, La Salle, Illinois, p. 76.

Pegge, C. D. (1962). The mode of the dream. *Journal of Mental Science* 108: 26–36.

Plato, The Cave – The Third Stage (515 e 5–516 e 2): The Genuine Liberation of Man to the Primordial Light – Part 1, Plato, Plato's Cave – the Second Stage, from The Republic, book VII, 514a–517a. taken from Platonis Oprea, recogn. Ioannes Burnet, Oxonii: clarendon, 2nd Edition (1905–1910) Vol. 4, (English trans.) M. Heidegger included in The Essence of Truth (2002) Continuum, London.

Rodowick, D. N. (1997). *Giles Deleuze's Time Machine*. Duke University Press, Durham, North Carolina.

Spoto, D. (1984). *The Dark Side of Genius: The Life of Alfred Hitchcock*. Ballentine, New York, p. 440.

Turim, M. (1989). *Flashbacks in Film: Memory and History*. Routledge, New York.

Ullman, M. and Limmer, C. (eds.) (1988). *The Variety of Dream Experience*. Continuum, New York.

Wilson, G. (1986). *Narration in Light: Studies in Cinematic Point of View.* Johns Hopkins University Press, Baltimore, Maryland, p. 2.

Zizek, S. (1992). Alfred Hitchcock, or, the form and its historical mediation. In *Everything You Always Wanted to Know About Lacan (But Were Afraid to Ask Hitchcock)*, ed. S. Lizek. Verso, London, pp. 1–14.

SECTION 6

Domhoff, G. W. (1993). The repetition of dreams and dream elements: A possible clue to the function of dreams. In *The Functions of Dreaming*, eds. A. Moffitt, M. Kramer and R. Hoffman. State University of New York Press, Albany, NY, pp. 293–320.

Fishbein, H. (1976). *Evolution, Development and Children's Learning.* Goodyear Publishing co. Inc., Pacific Palisades, CA, p. 8.

McManus, J., Laughlin, D. and Shearer, D. (1993). The function of dreaming in the cycles of cognition: A biogenic structural account. In *The Functions of Dreaming*, eds. A. Moffittt, M. Kramer and R. Hoffman. State University of New York Press, Albany, NY, pp. 21–50.

Pagel, J. and Broyles, K. (1996). Dream use in the creative process by screenwriters and directors. *ASD Abstracts.* p. 142.

Pagel, J. F. and Vann, B. (1992). The effects of dreaming on awake behavior. *Dreaming* 2(4): 229–237.

Pagel, J. F. and Vann, B. (1993). Cross-cultural dream use in Hawaii. *Hawaii Medical Journal* 52(2): 44–45.

Pagel, J. F., Vann, B. and Altomare, C. (1995). The association between stress and dreaming: Community background levels and changes with disaster (Hurricane Iniki). *Dreaming* 5(1): 43–50.

Pagel, J. F., Kwiatkowski, C. F. and Broyles, K. (2003). Creativity and dreaming: Correlation of reported dream incorporation into awake behavior with level and type of creative interest. *Creativity Research Journal* 15(2&3): 199–205.

CHAPTER 17

Bellak, L. (1958). Creativity: Some random notes to a systemic consideration. *Journal of Projective Techniques* 4: 363–380.

Bronowski, J.(1978). *The Origins of Knowledge and Imagination.* Yale Press, Hartford, Connecticut.

Bryon, L. Childe Harold's Pilgrimage, Canto III, Stanza 77, lines 726–732, *Lord Bryon: The Complete Poetical Works*, ed. Jerome J. McGann, Oxford University Press, Oxford, UK. Vol. 2, p. 105.

Bryon, L. (1969). *Lady Blessington's Conversations of Lord Bryon*, ed. E. Lovell. Princeton University Press, Princeton, NJ, p. 115.

Csikszentmihalyi, M. (1996). *Creativity: Flow and the Psychology of Discovery and Invention.* Harper Collins, New York.

Dasgupta, S. (1996). *Technology and Creativity.* Oxford University Press, New York.

Delaney, G. (1979). *Living Your Dreams.* Harper Books, San Fransisco, CA.

Finke, R. A., Ward, T. B. and Smith, S. M. (1992). *Creative Cognition: Theory Research and Applications.* The MIT Press, Cambridge, MA.

Freud, S. (1908/1959). Creative writers and day-dreaming. In *Standard Edition of the complete Works of Sigmund Freud*, Vol. 9, ed. J. Strachey. Hogarth Press, London.

Freud, S. (1925/1973). An autobiographical study. In *The Standard Editions of the Complete Psychological Works of Sigmund Freud*, Vol. XIV, ed. J. Strachey. Hogarth Press, London.

Freud, S. (1963) (1915–1916). *Introductory Lectures in Psychoanalysis*, Vol. XV, Standard Edition, ed. J. Hogarth press. Strachey. London, p. 17.

Garcia, J. D. (1991). *Creative Transformation: A Practical Guide for Maximizing Creativity*. Whitmore Publishing Co., Ardmore, PA.

Gardner, H. (1973). *The Arts and Human Development*. Basic Books, New York.

Gedo, J. (1983). *Portraits of the Artist*. Guilford, New York.

George Bernard Shaw (1972). *Back to Methuselah. Definitive Edition Bernard Shaw, Collected Plays with Their Prefaces*, Vol. 5. Dodd Mead and Co., New York, pp. 340–564.

Getzels, J. W. and Csikszentmihalyi, M. (1976). On the roles, values, and performance of future artists: A conceptual and empirical exploration. *Sociological Quarterly* 9: 516–550.

Ghiselin, B. (1952). *The Creative Process*. University of California Press, Berkeley, CA.

Gruber, H. E. (1981). *Darwin on Man: A Psychological Study of Scientific Creativity*, 2nd Edition. University of Chicago Press, Chicago, IL (original work published in 1974).

Guilford, J. P. (1959). Traits of creativity. In *Creativity and Its Cultivation*, ed. H. H. Anderson. Harper and Row, New York.

Guilford, J. P. (1965). A psychometric approach to creativity. In *Creativity in Childhood and Adolescence: A Diversity of Approaches* ed. H. H. Anderson. Science and Behavior Books, Palo Alto, CA.

Guilford, J. P. (1967). *The Nature of Human Intelligence*. McGraw-Hill, New York.

Harman, W. and Rheingold, H. (1984). *Higher Creativity: Liberating the Unconscious for Breakthrough Insights*. Jeremy P. Tarcher Inc., Los Angeles.

Hofstadter, D. (1985). *Metamagical Themas*. Basic Books, New York.

Jamison, K. (1993). *Touched with Fire: Manic-depressive Illness and the Artistic Temperament*. Simon and Schuster, New York.

John-Steiner, V. (1985). *Notebooks of the Mind*. University of New Mexico Press, Albuquerque, New Mexico.

Jung, C. (1923). On the relation of analytic psychology to poetic art. In *The Creativity Question*, eds. A. Rothenberg and Hausman. Duke University Press, Durham, NC (1976).

Jung, C. (1965). *Memories, Dreams, Reflections*. Vintage, New York.

Kavaler-Alder, S. (2000). *The Compulsion to Create – Women Writers and Their Demon Lovers*. Other Press, New York.

Keep, O. A. (1957). Discussant in the Arden House conference on motivating the creative process. In *The Nature of Creativity: Contemporary Psychological Perspectives*, ed. R. J. Sternberg (1991). Cambridge University Press, Cambridge, UK.

Kessler, R. C., McGonagle, K. C. and Zhao, S. (1994). Epidemiology of psychiatric disorders. *Archives of General Psychiatry* 51: 8–19.

Kligerman, C. (1980). *Art and the Self of the Artist, Advances in Psychology*, ed. A. Goldberg. International University Press, New York, pp. 383–396.

Kubie, L. (1958). *Neurotic Distortion of the Creative Process*. University of Kansas Press, Lawrence, Kansas.

Lee, H. (1940). A theory concerning free creation in the inventive arts. *Psychiatry*, 3.

Mannoni, M. (1999). *Separation and Creativity – Refinding the Lost Language of Childhood*, trans. S. Fairfield. Other Press, New York.

Martindale, C. (1990). *The Clockwork Muse: The Predictability of Artistic Change*. Basic Books, New York.

May, R. (1975). *The Courage to Create*. W. W. Norton, New York.

Melville, H. (1851). *Moby Dick: Or the Whale* (Reprint [1979]). University of California Press, Berkeley, CA.

Miller, A. (1988). *The Untouched Key: Tracing Childhood Trauma in Creativity and Destructiveness*. Anchor Books/Doubleday, New York.

Moliner, E. R. (1994). *The Conscious State of Matter*. Vantage Press, New York.

Random House Dictionary of the English Language – 2nd Edition Unabridged (1987), Editor-in-chief J. Stein. Random House Inc., New York.

Rogers, C. R. (1954). Toward a theory of creativity, ETC: A review of general semantics reprinted. In *Creativity and Its Cultivation*, Vol. 11, ed. H. H. Anderson. Harper and Row, New York.

Rothenberg, A. (1990). *Creativity and Madness*. The Johns Hopkins University Press, Baltimore.

Sternberg, R. J. (ed.) (1991). *The Nature of Creativity: Contemporary Psychological Perspectives*. Cambridge University Press, Cambridge, UK.

Tyrell, G. N. M. (1947). *The Personality of Man*. Penguin Book, London.

Wallas, G. (1926). *The Art of Thought*. Harcourt, Brace, New York.

Websters New Universal Unabridged Dictionary (1996). Barnes and Nobel Books, New York.

Wertheimer, M. (1945). *Productive Thinking*. Harper, New York.

West, T. G. (1991). *In the Minds? Eye*. Prometheus Books, Buffalo, NY.

CHAPTER 18

Bedard, M., Montplaisir, J., Richer, F. and Malo, J. (1991). Nocturnal hypoxaemia as a determinate of vigilance impairment in sleep apnea syndrome. *Chest* 100(2): 367–370.

Duffy, J. D. and Campbell, J. J. (1994). The regional pre-frontal syndromes: A theoretical and clinical overview. *Journal of Neuropsychiatry* 6(4): 379–387.

Fuster, J. M. (1989). *The Prefrontal Cortex – Anatomy, Physiology, and Neuropsychology of the Frontal Lobe*. Raven Press, New York.

Hartmann, E. (1991). *Boundaries of the Mind*. Basic Books, New York.

Jung, C. (1965). *Memories, Dreams, Reflections*. Vintage, New York.

Pagel, J. F. (2003). Non-dreamers. *Sleep Medicine* 4: 235–241.

Pagel, J. F. and Kwiatkowski, C. F. (2003). Creativity and dreaming: Correlation of reported dream incorporation into awake behavior with level and type of creative interest. *Creativity Research Journal* 15(2&3): 199–205.

Pagel, J. F. and Vann, B. (1992). The effects of dreaming on awake behavior. *Dreaming* 2(4): 229–237.

Pagel, J. F., Kwiatkowski, C. and Broyles, K. (1999). Dream use in film making. *Dreaming* 9(4): 247–296.

Plato, The Cave – The Third Stage (515 e 5–516 e 2): The Genuine Liberation of Man to the Primordial Light – Part 2, Plato, Plato's Cave – the Second Stage, from The Republic, book VII, 514a–517a. taken from Platonis Oprea, recogn. Ioannes Burnet, Oxonii: clarendon, 2nd Edition (1905–1910) Vol. 4 (English trans.) M. Heidegger included in The Essence of Truth (2002) Continuum, London.

Sands, S. and Levin, R. (1997). The effects of music on dream content: An empiric analysis. *Dreaming* 7(3): 215–220.

CHAPTER 19

Chandler, C. (1995). *I, Fellini*. Random House, New York.

Delaney, G. (1979). *Living Your Dreams*. Harper Books, San Francisco, CA.

Epel, N. (1993). *Writers Dreaming: Twenty-Six Writers Talk about Their Dreams and the Creative Process*. Vintage Books, New York.

Gardner, H. (1973). *The Arts and Human Development*. Basic Books, New York.

Kinder, M. (1988). The dialectic of dreams and theatre in the films of Ingmar Bergman. *Dreamwork* 5: 186.

Kurosawa, A. (1982). *Akira Kurosawa: Something Like an Autobiography*. Vintage Books, New York.

Magarshack, D. (1961). *Stanislavsky on the Art of Stage*. Hill and Wang, New York.

Moreno, Z. T. (1974). A survey of psychodramatic techniques. In *Psychodrama: Theory and Therapy*, ed. I. A. Greenberg. Behavioral Publications, New York, pp. 85–100.

Pagel, J. and Broyles, K. (1996). Dream use in the creative process by screenwriters and directors. *ASD Abstracts*. pp. 142–144.

Pagel, J. F. and Kwiatkowski, C. F. (2003). Creativity and dreaming: Correlation of reported dream incorporation into awake behavior with level and type of creative interest. *Creativity Research Journal* 15(2&3): 199–205.

Pagel, J. F. and Vann, B. (1992). The effects of dreaming on awake behavior. *Dreaming* 2(4): 229–237.

Pagel, J. F., Kwiatkowski, C. and Broyles, K. (1999a). Dream use in film making. *Dreaming* 9(4): 247–296.

Pagel, J. F., Kwiakowski, C. and Broyles, K. (1999b). Dream use in filmmaking. *Sleep* 22(Suppl): S173.

Pagel, J. F., Crow, D. and Sayles, J. (2003). Filmed dreams: Cinematographic and story line characteristics of the cinematic dreamscapes of John Sayles. *Dreaming* 13(1): 43–48.

Sayles, J. (1987). *Thinking in Pictures*. Houghton Mifflin Company, Boston, MA.

States, B. O. (1993). *Dreaming and Storytelling*. Cornell University Press, Ithaca, NY.

Tedlock, B. (ed.) (1992). *Dreaming: Anthropological and Psychological Interpretations*. School of American Research Advance Seminar Series, Sante Fe, New Mexico.

CHAPTER 20

Adorno, T. (1970). Translated by W. Mitchell. Continuum International Publishing Group, London, UK. *Aesthetic Theory*.

Arnheim, R. (1974). *Entropy and Art: An Essay on Disorder and Order*. University of California Press, Berkeley, CA, p. 55.

Fishbein, H. (1976). *Evolution, Development and Children's Learning*. Goodyear Publishing co. Inc., Pacific Palisades, CA, p. 8.

Freud, S. (1917). The paths to symptom-formation, Part III general theory of the neuroses. In *Introductory Lectures on Psychoanalysis (1966)*, trans. and ed. J. Strachey. W. W. Norton, New York, p. 375.

Genette, J. (1997). L'euve de l'art, 2, La Revelation esthetique.

Hartmann, E. (1991). *Boundaries in the Mind*. Basic Books, NYC, NY, pp. 174–184.

Heller, A. and Feher, F. (1989). *Reconstructing Aesthetics*. Basil Blackwell, Oxford.

Hobson, J. A. (1988). *The Dreaming Brain – How the Brain Creates Both the Sense and Nonsense of Dreams*. Basic Books, New York.

Hunt, H. (1989). *The Multiplicity of Dreams – Memory, Imagination and Consciousness*. Yale University Press, New Haven, p. 14.

Husain, M. (1990). The relevance of Aristotle's principles for the concept of structure in Checkhosovague Structuralism. *Znakolog* 2: 135–147.

La Capra, D. (2000). Reflections on trauma, absence and loss. In *Whose Freud? The Place of Psychoanalysis in Contemporary Culture*. Yale University Press, New Haven, pp. 178–204.

Mannoni, M. (1999). *Separation and Creativity – Refinding the Lost Language of Childhood*, trans. S. Fairfield. Other Press, New York.

Matisse, H. (1978). Notes d' un peintre. *La Grande Revue* LII: 24.

Passmore, J. (1991). *Serious Art*. Open court, La Salle, Illinois, pp. 11, 16.

Pegge, C. D. (1962). The mode of the dream. *Journal of Mental Science* 108: 26–36.

Plato, Republic (Book 10), trans. and ed. K. Dover. Cambridge University Press, Cambridge.

Plato's Sophist (360 BC). *Plato's Sophist – The Professor of Wisdom (1996)*, trans. E. Brann, P. Kalkavage and E. Salem. Focus Publishing, Newburyport, p. 267.

Scarry, E. (1995). On vivacity: The difference between daydreaming and imagining-under-authorial-instruction. *Representations* 52: 1–26.

Shaw, G. (1905/1957). Man & Superman: A comedy and philosophy, Penguin, New York.

States, B. (1997). *Seeing in the Dark – Reflections on Dreams and Dreaming*. Yale University Press, New Haven, pp. 1, 2, 7, 153.

Zeki, S. (1999). *Inner Vision: An Exploration of Art and the Brain*. Oxford University Press, Oxford, p. 22.

SECTION 7

Chandler, M. (1988). Doubt and developing theories of mind. In *Developing Theories of Mind*, eds. J. Astington, P. Harris and D. Olson. Cambridge University Press, Cambridge, p. 411.

Churchland, P. (1986). *Neurophilosophy – Towards a Unified Science of the Mind-Brain*. The MIT Press, Cambridge, MA, pp. 3, 482.

Crick, F. (1979). Thinking about the brain. *Scientific American* 241: 219–232.

Da Costa, N. and French, S. (2003). *Science and Partial Truth – A Unitary Approach to Models and Scientific Reasoning*. Oxford University Press, Oxford, pp. 61–62.

Einstein, A. (1949/1970). Autobiographical notes. In *Albert Einstein: Philosopher–Scientist*, ed. P. Schilpp. The Library of Living Philosophers, Third Edition, Vol VII, Evanston, Ill, pp. 3–95.

Hardcastle, V. (1996). *How to Build a Theory in Cognitive Science*. State University of New York Press, Albany, NY.

Langholm, T. (1988). *Partiality, Truth and Persistence*. Center for the Study of Language and Information, Stanford.

Popper, K. (1935/1959). *The Logic of Discovery*. Hutchinson, London.

Ramon y Cajal S. (1909–1911). Histologie du Systeme Nerveux de l'Homme et des Vertebres (Vols. 1 and 2), trans. S. Azoulay. Maloine, Paris.

CHAPTER 21

Bernstein, R. (1983). *Beyond Objectivism and Relativism*. University of Pennsylvania Press, Philadelphia, PA.

Churchland, P. (1986). *Neurophilosophy – Towards a Unified Science of the Mind–Brain*. The MIT Press, Cambridge, MA, pp. 3, 450.

Crick, F. (1979). Thinking about the brain. *Scientific American* 241: 219–232.

Damasio, A. (1994). *Descartes' Error: Emotion, Reason and the Human Brain*. Avon Books Inc., New York.

Hippocrates (1891/1994) Hippocratic Corpus: Works by Hippocrates, trans. A. Francis, The Internet Classics Archive, Daniel C. Stevens, Web Atomics 1994–2000.

Hobson, J. (1988). *The Dreaming Brain*. Basic Books, New York, p. 204.

Hobson, J. (1994). *The Chemistry of Conscious States*. Little, Brown and Company, Boston, MA, p. 205.

Hobson, J., Pace-Schott, E. and Stickgold, R. (2003). Dreaming and the brain: Toward a cognitive neuroscience of conscious states. In *Sleep and Dreaming: Scientific Advances and Reconsiderations*, eds. E. Pace-Schott, M. Solms, M. Blagrove and S. Harnad. Cambridge University Press, Cambridge, pp. 1–50.

Kahn, D. and Hobson, J. (2005). State-dependent thinking: A comparison of waking and dreaming thought. *Conscious Cognition* 14(3): p. 429.

Kandle, E. R. (2000). The brain and behavior. In *Principles of Neural Science*, 4th Edition, eds. E. R. Kandel, J. H. Schwartz and T. M. Jessell. McGraw Hill, New York, 5–18.

McCarley, R. and Hobson, J. (1975). Neuronal excitability modulation over the sleep cycle: A structural and mathematical model. *Science* 189: 58–60.

Nielsen, T. (2003). A review of mentation in REM and NREM sleep: "Covert" REM sleep as a possible reconciliation of two opposing models. In *Sleep and Dreaming: Scientific Advances and Reconsiderations*, eds. E. Pace-Schott, M. Solms, M. Blagtove and S. Harnad, pp. 59–74.

Pace-Schott, E. (2003). Postscript: Recent findings on the neurobiology of sleep and dreaming. In *Sleep and Dreaming: Scientific Advances and Reconsiderations*, eds. E. Pace-Schott, M. Solms, M. Blagrove and S. Harnad. Cambridge University Press, Cambridge, pp. 335–350.

Pagel, J. F. (2005). Neurosignals – Incorporating CNS electrophysiology into cognitive process. *Behavior and Brain Science* 28(1): 75–76.

Pagel, J. F., Dawson, D., Drozd, J. and Snyder, S. (2006). Quantitative EEG (QEEG) power characteristics of children with OSA in a pediatric AD/HD population. *Sleep* 29(Abstract supplement): A337–A338.

Rojas, D., Teal, P., Sheedeer, J. and Reite, M. (2000). Neuromagnetic alpha suppression during an auditory Sternberg task. Evidence for a serial, self-terminating search of short-term memory. *Brain Research Cognitive Brain Research* 1091–1092; 85–89.

Rojas, D., Maharajh, K., Teale, P., Kleman, M., Benkers, T., Carlson, J. and Reite, M. (2006). Development of the 40 Hz steady state auditory evoked magnetic field from ages 5 to 52. *Clinical Neurophysiology* 117(1): 110–117.

Rossi, E. (2002). The psychobiology of gene expression: Neuroscience and neurogenesis. In *Hypnosis and the Healing Arts*. W. W. Norton, New York, pp. 12–17.

Ryle, G. (1949). *The Concept of Mind*, New University Edition. University of Chicago Press, Chicago, IL.

Schwartz, J. H. (2000). Neurotransmitters. In *Principles of Neural Science*, 4th Edition, eds. E. R. Kandel, J. H. Schwartz and T. M. Jessell. McGraw Hill, New York, pp. 280–297.

Searle, J. (1984). *Minds, Brains and Science*. Harvard University Press, Cambridge, MA.

Young, J. (1987). *Philosophy and the Brain*. Oxford University Press, Oxford.

CHAPTER 22

America's Mental Health Survey (2001). National Mental Health Association.

Bechtel, W., Mandik, P. and Mundale, J. (2001). *Philosophy Meets the Neurosciences in Philosophy and the Neurosciences – A Reader*, eds. W. Bechtel, P. Mandik, J. Mundale and R. Stufflebean. Blackwell Publishers, Oxford, p. 6.

Descartes, R. (1641). Objections against the meditations and replies. In *Great Books of the Western World: Bacon, Descartes and Spinoza (1993)*, Editor-in-Chief M. J. Adler. Encyclopaedia Britannica, Inc., Chicago, IL, p. 100.

DSM-IV (1994). *Diagnostic and Statistical Manual of Mental Disorders*, 4th Edition, Francis A. (Chair – Task Force on DSM-IV). American Psychiatric Association, Washington, D.C.

Foulkes, D. (1985). *Dreaming: A Cognitive–Psychological Analysis*. Lawrence Erlbaum, Hillsdale, NJ.

Freud, S. (1917a). Psychoanalysis and psychiatry general theory of the neuroses. In *Introductory Lectures on Psychoanalysis (1966)*, trans. and ed. J. Strachey. W. W. Norton, New York, p. 255.

Freud, S. (1917b). Some analyses of simple dreams, general theory of the neuroses. In *Introductory Lectures on Psychoanalysis (1966)*, trans. and ed. J. Strachey. W. W. Norton, New York, p. 184.

Freud, S. (1923). The ego and the id, trans. 1927, London; reprinted In *Complete Psychological Works*, ed. J. Strachey. Hogarth Press, London. Vol. XIX (1961).

Freud, S. (1925/1973). An autobiographical study. In *The Standard Editions of the Complete Psychological Works of Sigmund Freud*, Vol. XIV, ed. J. Strachey. Hogarth Press, London.

Freud, S. (1933/1973). *New Introductory Lectures on Psychoanalysis*. Penguin, Hammondsworth, pp. 83, 193.

Frosh, S. (1998). *Psychoanalysis, Science and Truth in Freud 2000*, ed. A. Elliott. Routledge, NewYork, p. 36.

Grunbaum, A. (1984). *The Foundations of Psychoanalysis: A Philosophical Critique*. University of California Press, Berkeley, CA.

Hunt, H. (1989). *The Multiplicity of Dreams: Memory, Imagination and Consciousness*. Yale University Press, New Haven, p. 14.

Jung, C. (1953–1971). *The Collected Works*. London.

Kaplan, A. (ed.) (1990). *Psychoanalysis and Cinema*. Routledge, New York, p. 7.

States, B. O. (1993). *Dreaming and Storytelling*. Cornell University Press, Ithaca, NY.

Wallerstein, R. (1986). *Forty-Two Lives in Treatment*. Guilford, NY, pp. 304–305.

CHAPTER 23

Heidegger, M. (1988). *The Essence of Truth – On Plato's cave Allegory and Theaetetus*, trans. T. Sadler. Continuum, London, p. 54.

McGinn, C. (1991). *The Problem of Consciousness*. Blackwell, Oxford, p. 5, 9.

Plato's Cave – The Fourth Stage – the Freed Prisoner's Return to the Cave from The Republic Book VII (516 e 3–517 a 6), Plato, Plato's Cave – the Second Stage, from The Republic, book VII, 514a–517a. taken from Platonis Oprea, recogn. Ioannes Burnet, Oxonii: clarendon, 2nd Edition (1905–10) Vol. 4 (English trans.) M. Heidegger included in The Essence of Truth (2002) Continuum, London.

CHAPTER 24

Descartes, R. *Discourse on Method (1637) and Meditations on First Philosophy (1641)*, trans. D. A. Cress (1980). Hackett Publishing Company, Indianapolis, In.

McGinn, C. (1989). Can we solve the mind–body problem? *Mind* 98: 349–350, 358.

Scarry, E. (1995). On vivacity: The difference between daydreaming and imagining-under-authorial-instruction. *Representations* 52: 1–26.

Young, J. (1987). *Philosophy and the Brain*. Oxford University Press, Oxford.

INDEX